The Australia

House Building Manual

Pinedale Press

Other Books by the Same Author:
Take Me Back to the Eighties
  (A Pen and Ink Sketch Book of 19th
  Century Australian Homes & Buildings)
The New Zealand Owner Builders Manual
The Australian Owner Builders Manual
How to be a Successful Owner Builder &
  Renovator
The Australian Renovators Manual
The Australian Roof Building Manual
  (Hiddle & Staines)
House Building the Easy Hebel Way
Australian Decks & Pergolas Construction Manual

Published by Pinedale Press

National Library of Australia
Cataloguing-in-publication data

Staines, Allan 1942
National Library of Australia card number and
ISBN: 978-187521730-4

Holder of Copyright
Pinedale Press
2 Lethbridge Court
CALOUNDRA QLD 4551
AUSTRALIA
For Orders, please fax (07) 5491 9219

1st Edition	September	1998
Reprint	December	1998
Reprint	June	1999
Reprint	May	2000
2nd Edition	May	2001
Reprint	January	2002
3rd Edition	March	2003
Reprint	September	2003
4th Edition	March	2004
Reprint	November	2004
Reprint	August	2005
Reprint	March	2006
5th Edition	January	2007
Reprint	August	2007
6th Edition	March	2010

Acknowledgements
*The Author and Publisher wish to express appreciation to the
following organisations and individuals for their contributions,
proof reading and constructive criticism:* Colin E. MacKenzie,
engineer Timber Qld., Timber Development Association of S.A.,
Clay Brick and Paver Institute of NSW, Qld & WA and Clay Brick
Engineer Specialist—Dr. Stephen Lawrence ME, PhD, FIEAust,
CPEng, NPER (structual) of SPL Consulting. Peter Bayetto
BE, FIE Aust., CPE ng R.P.E.Q, Block Masonry Engineer-Rod
Johnston.B Tech, M Eng, Sc, MIC D, MIE Aust., Brickwork
Engineer. Cement, Concrete & Aggregates Aust., CSR Building
Products, James Hardie Building Boards, G. James Glass &
Aluminium, Pryda Australia, Besser Masonry, Abey Australia,
Concrete Masonry Assn. of Aust., Carter Holt Harvey, Bostik
Australia, Selleys Chemical Co., Jesper Jensen and Anthony
Milostic of James Hardie.

To Australian Building Codes Board for their kind permission to
use their table on general classifications of sites from the BCA as
well as portions of their M.S. Lintel table.

*To the following individuals for their constructive criticism
and input:* Bill Moore P.C. Eng, BE, M.I.E Aust. R.P.E.Q2403.
Practising Architect and Lecturer-Charles Ham A.R.A.I.A, Building
materials construction Consultant-Lex Somersville, Brian Peach of
Peach Institute, Building Surveyor-Steven Tucker BBS, National
Trainer to the Thin Section Rendered Coatings Industry-Scott
Squire and overall proof reading Charlie Herbert.

To Noela Ollenburg and Ryan Aldous Graphic Design for their
computer skills in preparing this material and to Paul Cooper for
his penmanship over the author's drawings.

Contents

Publisher's Notes & Instructions

About this Manual
This manual is designed as an instructional guide for builders, designers, apprentices, cadets and owner builders. It covers the four conventional house building systems used in Australia — **Brick veneer**, **Timber Frame**, **Cavity brick** and **Concrete block**.

While the 'BCA' and 'Australian Standards' provide the legal building requirements, this manual is intended to explain the why and how to put them into practice. The graphical interpretation and step-by-step description is intended for easy comprehension by the apprentice or cadet.

This manual is intended as a general guide to common on-site building methods. Specific dimensions, specifications and design of structural members must be obtained from individually prepared plans and specifications.

Methods Adopted in this Manual
Methods and details have been chosen through continued consultation with engineers and advisors in national industry, as well as the Timber, Concrete, Concrete block and Clay Brick Associations. While many variations on a particular detail or method may exist, those represented have been selected for their structural soundness, weatherproof qualities and because they are generally approved of by engineers in industry and in accordance with the 'Building Code of Australia'.

BCA & Australian Standards Codes
This manual highlights only common on-site building situations and is not a building code. While every effort has been made to conform to the requirements of the 'BCA' and 'Australian Standards' these codes are under constant review and change.
Should doubt arise regarding a particular situation, reference should be made to the most recent edition or amendment of these codes.

Proof Readers
Proof readers are selected from engineers and professionals who are held in high regard by the relevant associations and who possess a sound working knowledge of relevant building legislation.

Manufacturer's Specifications
This term refers to specifications or fixing instruction booklets or pamphlets produced by manufacturers. It is important to read these instructions as warranties are often dependant upon installation being carried out in accordance with them.

Soil Testing, Footing & Slab Design
Future structural problems may result if footings are *not* designed to suit each particular building site. Variations in sub-soils can occur even between adjoining sites. It is advisable to have designs and subsoil tests for footings and slab floors performed by a competent engineer.

Site Inspections
Where site and construction inspections are *not* a requirement, future structural failure may become the builder's legal responsibility. For this reason, it is good practice to have an engineer or building surveyor carry out inspections of footings, slab and frame or masonry work on cavity brick or concrete block houses, followed by a written certification that the work was constructed according to the approved plans and specifications. These certificates may also enhance the future resale of the house.

Roof Framing
This manual includes erection and construction of manufactured roof trusses and near flat roofs. For 'How to Pitch the Conventional Cut On-Site Roof' and for roof tables and bevels refer to *'The Roof Building Manual'*.

Advice for Owner Builders
Many homes are built by owner builders, a large number of whom are carpenters, bricklayers and other building tradespeople. However, some owner builders are inexperienced in carpentry. For these people it will be beneficial to engage a carpenter to work alongside or have a Building Advisor regularly visit the site. The added cost will be more than compensated for through savings in time and materials.

Explanation of Design Wind Speed
The design of timber framing, brickwork and other structural members is closely related to the design wind speed allocated to the individual site.

The average wind velocity for each locality has been established over a 50 year period and then given the following classifications. 'N' represents Non-cyclonic and 'C' represents Cyclonic. *The categories are:* 'N1' with a design gust wind speed of 34m/sec (metres per second), 'N2' = 40m/sec, 'N3' and 'C1' indicated as N3/C1 = 50m/sec, N4/C2 = 61m/sec, N5/C3 = 74m/sec and N6/C4 = 86m/sec.

To determine the design wind speed for an individual site enquire at either the local government building dept, a local designer, architect or engineer.

Typical Construction Methods

This chapter will deal with typical details and methods commonly adopted for each of the four traditional building systems outlined below. The drawings below are enlarged on the following pages.

TIMBER FRAME (Page 6)

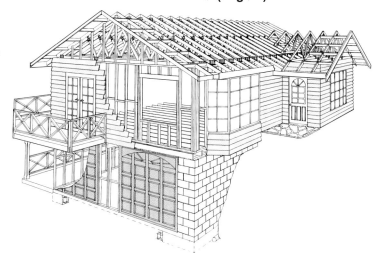

The timber framed house can be applied to one or two storey. The external surface can be clad with timber or manufactured weatherboards and panelling, or acrylic render can be applied. Structural load bearing is performed by the timber frame.

BRICK VENEER (Page 11)

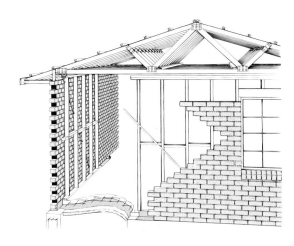

The brick walls are a facade anchored with brick ties to a timber or steel frame.The frame performs the structural load bearing work supporting the roof, ceiling and wall linings. A cavity space is provided between the frame and brick wall. This cavity acts as a moisture barrier.

CAVITY BRICK (double brick) (Page 26)

Cavity brick construction is two brick walls standing side by side separated by a cavity, tied together with brick ties that cross the cavity and embedded into the mortar joints. Floors can be either concrete slab or timber.

HOLLOW CONCRETE BLOCK MASONRY (190 single leaf) (Page 36)

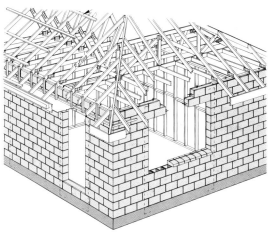

Reinforcing and grout is usually applied to specific vertical cores, lintels and a top perimeter bond beam. Walls can be cement or acrylic rendered on one or both sides or simply painted to reduce costs. Alternatively, plasterboard linings can be applied to the internal surface.

The Timber Frame House

This system of house building is used where light weight cladding such as weatherboards are attached to the frame. The frame system is also used as the structural load bearing shell of all brick veneer houses. The walls can be constructed on a concrete slab (the simplest means) or over a half or full basement as illustrated using a bearer and joist flooring system. Natural tongue and groove timber, particle board, plywood sheet or compressed fibre cement sheet flooring is laid. Termite protection should be provided in accordance with the A.S. Termite Management Code.

This application is carried out by a Solid Plasterer. Decorative textures can be sprayed over the surface then painted with a quality 100% acrylic paint system.

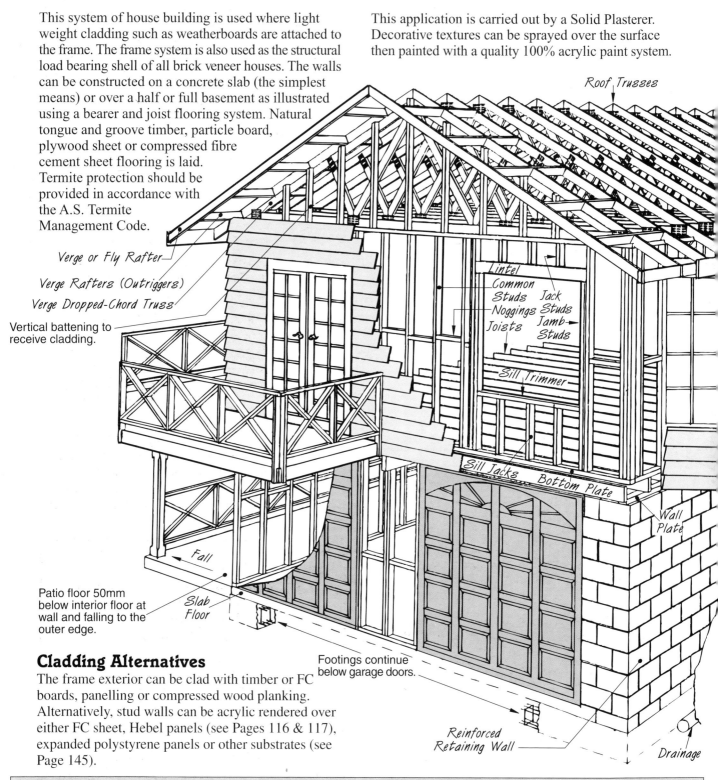

Roof Trusses

Verge or Fly Rafter

Verge Rafters (Outriggers)

Verge Dropped-Chord Truss

Vertical battening to receive cladding.

Lintel
Common Studs
Jack Studs
Noggings
Jamb Studs
Joists
Sill Trimmer

Sill Jacks Bottom Plate

Wall Plate

Fall

Patio floor 50mm below interior floor at wall and falling to the outer edge.

Slab Floor

Footings continue below garage doors.

Reinforced Retaining Wall

Drainage

Cladding Alternatives

The frame exterior can be clad with timber or FC boards, panelling or compressed wood planking. Alternatively, stud walls can be acrylic rendered over either FC sheet, Hebel panels (see Pages 116 & 117), expanded polystyrene panels or other substrates (see Page 145).

Stages of Construction

1. Foundation, and footings, basement walls and piers are constructed as well as sanitary drainage under slab.
2. Concrete floors or bearers, joists and floor is laid (if a platform is being used).
3. Wall frames are erected, plumbed and braced true and straight.
4. Roof framing is constructed including fascia and barge boards fitted.
5. Roof sarking is applied and cladding attached.
6. Windows and exterior doors are installed.
7. External wall cladding applied and soffits lined.
8. Tongue and groove strip flooring is laid if specified.
9. Ceilings battened and/or noggings installed.
10. Plumbing pipes and electrical wiring are installed.
11. Internal ceilings and walls lined.
12. Bath, shower and kitchen cupboards are installed.
13. Internal doors and all mouldings are fitted.
14. Plumbing finished off.
15. Painting and decorating is carried out.
16. Floors are sanded if required.
17. Electrical work is finished off.

The Structural Design of Members

Timber framing members used in house framing are selected and designed according to the vertical and side loads they are required to support or resist such as walls, floor, roof and ceilings loads as well as horizontal wind loads. *Additional selection criteria could include:* Availability, cost, cladding to be applied, and the builder's personal preference for a particular species. The design of the cross sectional size, spacing, anchorage and bracing of members can be found in the 'AS 1684.2.

House plans should contain all of this information. House plans and specifications are legal documents to protect the builder, owner and subcontractors. To avoid future disputes and structural failure, it is important that the builder ensure these structural specifications are adhered to. *For example,* if the plans specify 'F14 graded hardwood' for wall frames, this can't be altered without consulting the designer and building surveyor.

Load Bearing & Non Load Bearing

Internal non-load bearing walls are generally referred to as partitions. Walls supporting the roof and ceiling loads are termed load bearing walls. Non-load bearing walls or partitions carry *no* roof or floor loads but may support ceiling loads only. These may also contain bracing.

Alternative Framing Methods

Framing Can be Carried out in one of Three Ways:

1. Factory pre-cut and pre-nailed.

2. Factory pre-cut and assembled on-site.

3. Wholly cut out and assembled on-site.

Each should be individually considered and costed. Where time is at a premium, it may be advisable to opt for pre-cut and pre-nailed frames.

Timber Species for House Frames

Radiata, Slash Pine & other Plantation Softwoods

These timbers are easy to saw and nail, are lightweight and dimensionally stable when seasoned.

NAIL FASTENINGS for structural steel connector plates, straps, braces and veneer ties are manufactured so that the heads will not easily shear off. While the B.C.A and AS may refer to them as 'flat head nails' manufacturers use various other terms such as: Pryda nails, Teco nails, Connector nails or Strapping nails. *Never substitute these with clouts.* While the AS requires these nails to be 30mm x 2.8dia., it is preferable to use 3.15dia. as these fit the manufactured hole diameters more closely.

NOMINAL NAILING VERSUS SPECIFIC AS 1684.2 Parts 2&3 Clause 9:6:1 states, *'Unless otherwise specified the specific details and/or fastenings required are in addition to nominal nailing, see Clause 9.5'.* This means the specific fastening is required as well as nominal. The specific *doesn't* necessarily replace the nominal.

Softwoods are more readily available. When nail fastening, use flat head nails to prevent pull-through.

Hardwood Although generally heavier than pine and harder to nail and saw, its higher strength enables smaller sectional sizes to be utilised in some instances. Nailing of members e.g. joists etc, often requires predrilling to avoid splitting when fastening. Splintering can be a problem. When nail fastening members, use bullet head nails or mechanically driven nails.

Cypress This has the least shrinkage problem of all framing timbers and is resistant to termite and borer attack. When nailing in a dry state, care must be taken to avoid splitting. Use blunt or shear point nails. Also when sawing etc, some tradespeople object to its strong aroma.

Listed below are actual timber framing sizes nominated for plantation pines as well as hardwood and Cypress. Some of these sizes may have to be ordered in advance. All timber lengths are purchased in 300mm increments.

PLANTATION PINES

70x35	70x45	70x70	90x90
90x35	90x45	90x70	
120x35	120x45	120x70	
140x35	140x45	140x70	
190x35	190x45	190x70	
240x35	240x45	240x70	
290x35	290x45	290x70	

HARDWOOD & CYPRESS

75x38	75x50	275x50	225x75
100x38	100x50	300x50	250x75
125x38	125x50	75x75	275x75
150x38	150x50	100x75	300x75
175x38	175x50	125x75	100x100
200x38	200x50	150x75	125x100
225x38	225x50	175x75	150x100
250x38	250x50	200x75	125x125

Termite Control *See Page 141 for details.*

Stress Grading

All structural framing members should be stress graded and approved plans should specify the specific grades to use.

Structural framing members are marked with their individual stress grade strength. The higher the stress grading number the greater the strength.

Gradings available are: F4, F5, F7, F8, F11, F14, F17, F22 and F27. In many situations the higher stress grade may enable the use of smaller sectional sizes.

Pine can be graded with an 'F' value or 'MGP' (Machine Graded Pine) followed by the MPa stress grading. MGP utilises a different and higher value system from 'F' grading. *These include:* MGP10, MGP12 and MGP15. When MGP grades are specified, they must *not* be substituted with 'F' graded timber without recalculating the sizes.

Seasoned or Unseasoned

Seasoning is the term applied to timber which has had its moisture content reduced by either kiln or air drying. Moisture quantity is expressed in percentages e.g. 12% moisture content.

Unseasoned and green timber is likely to shrink or suffer distortion (warping, bowing etc) after installation. This can cause serious problems particularly with brick veneer construction if members such as floor joists and bearers are unseasoned. The whole house can shrink down with the joists and bearers 10-20mm causing linings to crack and joinery to bind. It is advisable to use a non shrink flooring system such as hyJOISTS or similar.

Timbers for Verandahs & Outdoors

To select the right timber for use outdoors exposed to the weather or in-ground, you must consider the ability of the timber to resist decay and termites.

Timber treated to the appropriate 'H' level from AS 1604.2 for the specific end use is suitable. Timber of an in-ground Durability Class 1 is suitable for in-ground use while timber of an above-ground Durability Class 1 or 2 is suitable for above ground use. The use of untreated sapwood in an exposed situation should be avoided.

Cypress is suitable for decks and pergolas, however, for in-ground use, all sapwood must be removed.

Inground Applications

Where possible it is best to attach posts to engineered hot dip galvanised brackets or stirrups above ground. Many of the products available are *not* engineer designed for the purpose or for longevity. The brackets and stirrups are embedded into concrete in-ground. However, treated posts are often applied directly inground or embedded in concrete or crusher dust.

The end portion of treated posts embedded in-ground in concrete or crusher dust should not be trimmed on-site otherwise the untreated core of the post will then be exposed. Pressure treatments sometimes *don't* penetrate to the central core of the timber. Preservative coatings applied on-site *don't* provide as good a protection as pressure treatment.

'H' Treatment Levels The treatment level will depend on the intended purpose. For example, timber used in-ground will require a higher treatment level than for above ground timber such as hand railing or for where the timber will be given opportunity to dry out quickly after wet weather. The level of treatment is prefaced by the letter 'H' (the initial for the word 'Hazard').

H3 treated timber is used for above ground timbers **only** and H5 for inground. H4 is permissible in some states but is only suitable for moderate in-ground Hazard applications. If in doubt, use H5.

The table below provides a description of treatments and end uses.

Dry-After-Treatment or Seasoned After Treatment
Dry-after-treatment or seasoned after treatment for softwoods are preferred as they will be more stable and give less movement on-site. The timber is firstly kiln dried to 10-15% moisture content then put through the vacuum pressure treatment to receive the preservative then kiln dried again to 10-15% moisture content. Request 'Treated and Seasoned'.

Which Treatment Most common are CCA (Copper Chromium Arsenic), or ACQ (Ammoniacal Copper Quat), or Copper Azole. CCA is *not* permitted on decks, handrails, steps or human contact locations.

User Safety Precautions
The following safety precautions apply when using all preservative treated timbers.
 a). Wear gloves.
 b). Wash hands and face before eating, drinking or smoking.
 c). When sanding timber, provide good ventilation and wear protective clothing plus dust mask.
 d). NEVER USE IN FIRES FOR HEATING, COOKING OR B.B.Q.'s.
See also manufacturers' data sheets.

End Use	'H' TREATMENT LEVEL
ABOVE GROUND for Posts in stirrups, bearers, joists, ledgers, handrails, step stringers, step treads, bracing timber, deck flooring, pergola beams, joists & battens.	H3
IN GROUND and subject to severe wetting and leaching	H4
IN GROUND and subject to extreme wetting and leaching and/or where the critical use requires a higher degree of protection	H5
IN MARINE CONDITIONS Posts for decks or jetties	H6

Some Common Timber Framing Details

Further construction details can be found throughout the manual.

For aluminium window installation see Page 118.
For patio sliding doors see Page 119.

Joinery Installation

For window & door installation see Pages 118-122.

Some designers prefer to extend the top weatherboard above the soffit line as an additional protection should eaves become flooded from overflowing guttering. Wall battening *not* always being continuous, trapped water can seep across to the frame.

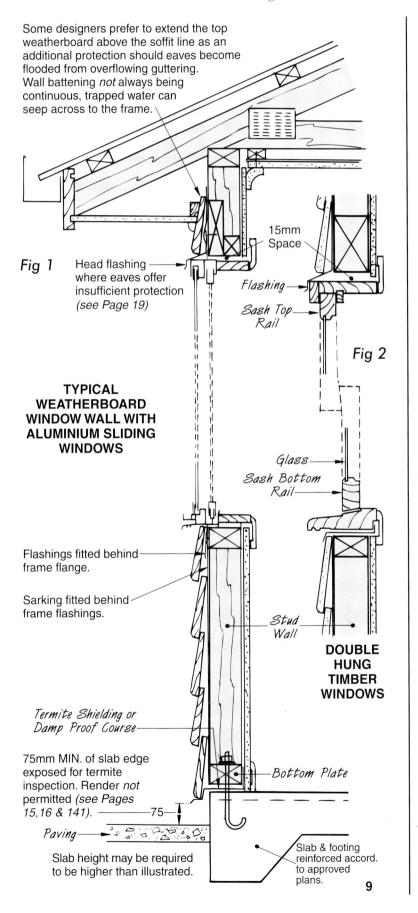

Fig 1

Head flashing where eaves offer insufficient protection *(see Page 19)*

15mm Space

Flashing

Sash Top Rail

Fig 2

TYPICAL WEATHERBOARD WINDOW WALL WITH ALUMINIUM SLIDING WINDOWS

Glass

Sash Bottom Rail

Flashings fitted behind frame flange.

Sarking fitted behind frame flashings.

Stud Wall

DOUBLE HUNG TIMBER WINDOWS

Termite Shielding or Damp Proof Course

75mm MIN. of slab edge exposed for termite inspection. Render *not* permitted *(see Pages 15, 16 & 141).* ——— 75

Paving

Bottom Plate

Slab height may be required to be higher than illustrated.

Slab & footing reinforced accord. to approved plans.

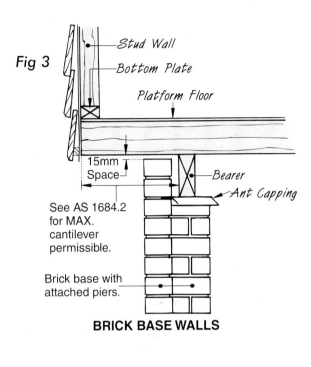

Fig 3

Stud Wall

Bottom Plate

Platform Floor

15mm Space

Bearer

Ant Capping

See AS 1684.2 for MAX. cantilever permissible.

Brick base with attached piers.

BRICK BASE WALLS

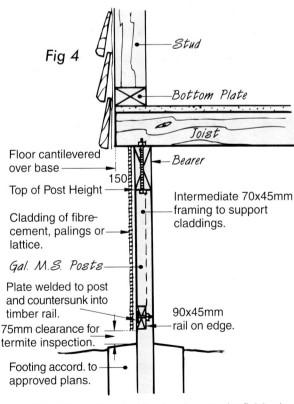

Fig 4

Stud

Bottom Plate

Joist

Floor cantilevered over base

Top of Post Height 150

Cladding of fibre-cement, palings or lattice.

Gal. M.S. Posts

Plate welded to post and countersunk into timber rail.

75mm clearance for termite inspection.

Footing accord. to approved plans.

Bearer

Intermediate 70x45mm framing to support claddings.

90x45mm rail on edge.

Should the rail or cladding be closer to the finished ground or paving surface than 75mm, treated or termite resistant timber such as cypress wall and roof framing should be used throughout.

POST TO FLOOR DETAIL

9

Timber Wall Framing 2-Storey Construction

Designing the House to Suit the Site

A sloping site can be utilised to include a half or full basement. Additional excavation may be necessary. Utilising the contour of the land in this way means that part of the house, whether it be garaging, laundry, rumpus, spare room or office or living areas, are located underneath. This can be a far more economical method of construction than having all rooms on a slab floor on the same level. Savings can be achieved as additional foundations and roof are *not* required for the rooms below.

Construction Method

2-Storey framing is best constructed using the platform method as illustrated in fig 1, with the first floor walls and floor being constructed, followed by the upper walls. 2-Storey wall frames can be clad as for single storey or brick veneer can be applied.

Painting the Top Storey of 2-Storeys

To save costs on reconstructing scaffolding, paint the upper storey exterior walls, soffits, fascia boards and guttering after installing the cladding.

> *Note:* Scaffolding is also required to the roof perimeter while roof battening and cladding are being installed. This applies to both single and 2-storey houses as a safety measure.

Stages of Constructing 2-Storeys

1. Foundations, footings and sanitary drainage under slab are constructed.
2. Basement floor is laid.
3. Basement wall frames with supporting beams erected.
4. Floor joists are laid (& flooring if a platform floor is used).
5. Upper level wall frames are erected.
6. Roof framing is constructed including fascia, barge boards and guttering.
7. Roof cladding is attached.
8. Windows and exterior doors are fitted.
9. Exterior cladding is attached.
10. Internal flooring is laid if a cut-in floor is used.
11. Plumbing and electrical are roughed-in.
12. Ceiling and wall linings are attached.
13. Bath, shower, kitchen cupboards etc are installed.
14. Doors are installed and all mouldings attached.
15. Wall and ceiling lining joints and fastener heads are filled and finished off.
16. Plumbing is finished off.
17. Wall and floor tiling is carried out.
18. Painting and decorating is carried out.
19. Electrical work is finished off.
20. Floor sanding (this can be done before decorating).

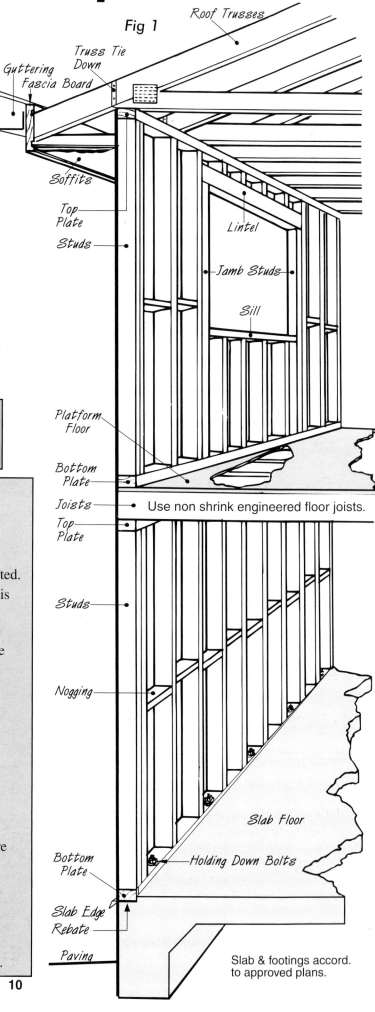

Fig 1

Roof Trusses
Truss Tie Down
Guttering
Fascia Board
Soffits
Top Plate
Studs
Lintel
Jamb Studs
Sill
Platform Floor
Bottom Plate
Joists
Use non shrink engineered floor joists.
Top Plate
Studs
Nogging
Slab Floor
Bottom Plate
Holding Down Bolts
Slab Edge Rebate
Paving
Slab & footings accord. to approved plans.

Brick Veneer Construction

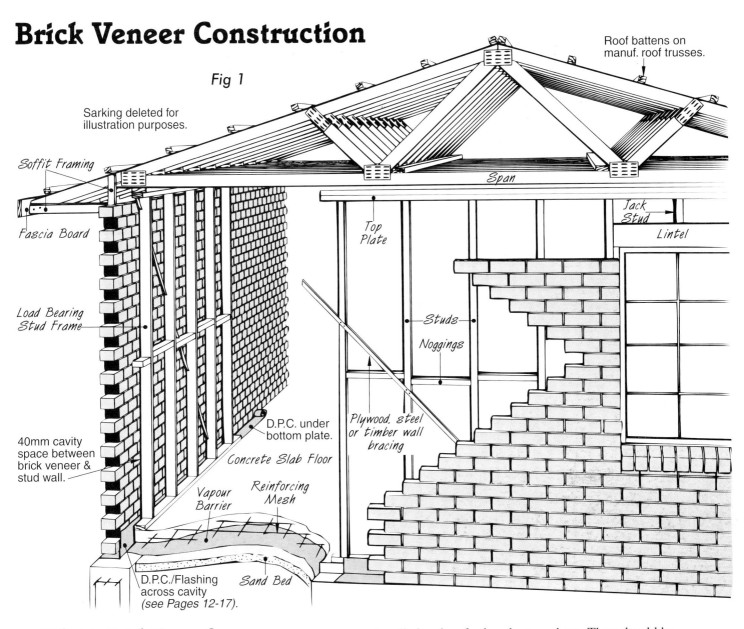

Fig 1

Roof battens on manuf. roof trusses.

Sarking deleted for illustration purposes.

Soffit Framing

Fascia Board

Load Bearing Stud Frame

40mm cavity space between brick veneer & stud wall.

D.P.C. under bottom plate.

Concrete Slab Floor

Vapour Barrier

Reinforcing Mesh

D.P.C./Flashing across cavity (see Pages 12-17).

Sand Bed

Top Plate

Span

Jack Stud

Lintel

Studs

Noggings

Plywood, steel or timber wall bracing

What is Brick Veneer?

Brick veneer is a brick facade fixed to a timber or steel frame with brick ties regularly spaced. The timber frame does all the structural load bearing work supporting the roof, ceiling and wall linings. An air space (cavity) of typically 40mm is provided between the frame and brick wall. This cavity acts also as a moisture barrier and provides some insulating properties.

Brick veneer can be utilised as single storey on concrete slab as in figs 1 & 2, or 2-Storey as in fig 1, Page 14. It can also be applied to the base portion of 2-Storeys with weatherboard or thin section render applied to the upper level as in fig 2, Page 14.

When is the Brick Wall Constructed?

Where brick courses are required below a slab floor or on houses with a timber floor, the bricks are laid up to the floor level. The remaining brickwork is laid after the roof is framed, windows, exterior door frames and meter box are installed.

Brick Delivery

The site should be signposted to ensure bricks are delivered close to the work and *not be* double

handled and so further damage them. They should be protected from rain to ensure they will be laid in a dry state.

Stages of Construction

1. Foundation, footings and sanitary drainage under slab.
2. Brick base constructed to floor level.
3. Floor laid, whether concrete or timber.
4. Timber wall frames are erected including bracing.
5. Roof framing is completed & tiles with sarking attached.
6. Windows and exterior door frames are installed including all flashings and sarking.
7. Brick walls are laid.
8. Soffits are framed and lined.
9. Electrical wiring and plumbing pipe out & drainage.
10. Ceilings and walls are lined.
11. Internal doors, bath, shower, kitchen cupboards and all mouldings are fitted.
12. Floor and wall tiling is carried out.
13. Plumbing and PC items are completed.
14. Painting and Decorating carried out.
15. Electrical fittings and switches are installed.
16. Floor sanded and carpets are laid.

Clay Brick Modules

The cutting of bricks should be limited to halves only. To achieve this the builder should ensure that horizontal and vertical steps in footings, and the placement of fitments such as doors and windows are dimensioned to suit these modules.

THE STANDARD METRIC BRICK MODULE

The length of a metric brick equals twice its width plus one joint. This varies from brick to brick. Brick joints are maintained at approximately 10mm. However, perpend joints are usually adjusted to limit the cutting of bricks to halves only.

Ensure cavity ties always fall to the external face and that the inner joints are closely aligned with the outer ones and never lower than the outer ones.

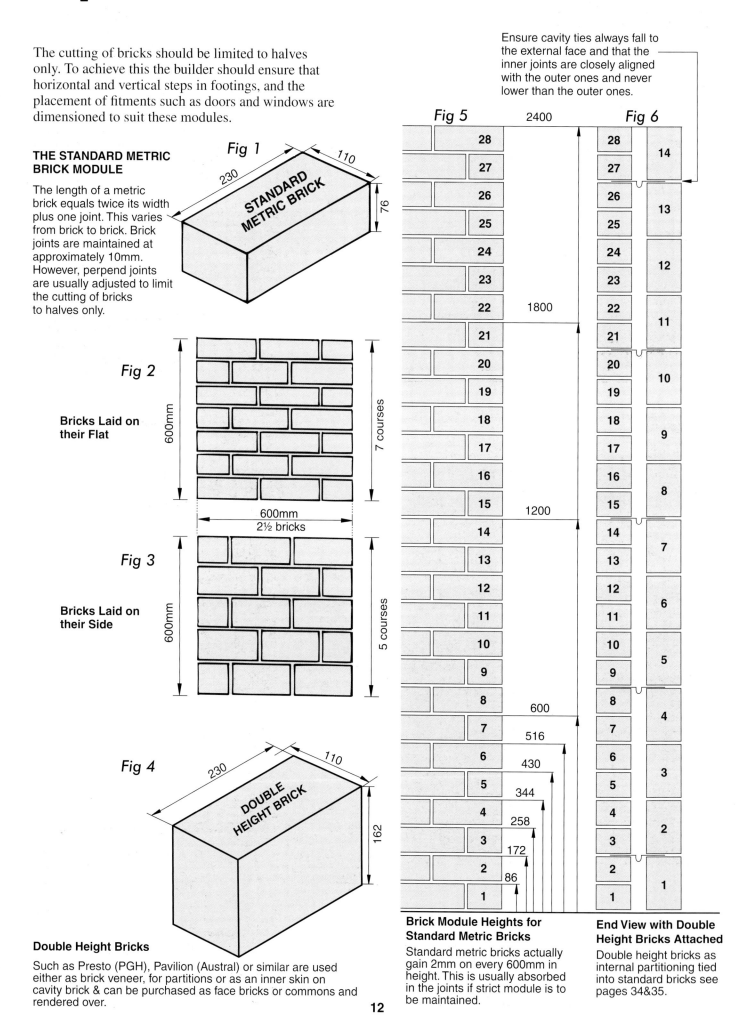

Fig 1
STANDARD METRIC BRICK — 230, 110, 76

Fig 2
Bricks Laid on their Flat — 600mm, 7 courses

Fig 3
Bricks Laid on their Side — 600mm, 600mm 2½ bricks, 5 courses

Fig 4
DOUBLE HEIGHT BRICK — 230, 110, 162

Fig 5 2400 ... 1800 ... 1200 ... 600, 516, 430, 344, 258, 172, 86

Fig 6

Double Height Bricks

Such as Presto (PGH), Pavilion (Austral) or similar are used either as brick veneer, for partitions or as an inner skin on cavity brick & can be purchased as face bricks or commons and rendered over.

Brick Module Heights for Standard Metric Bricks

Standard metric bricks actually gain 2mm on every 600mm in height. This is usually absorbed in the joints if strict module is to be maintained.

End View with Double Height Bricks Attached

Double height bricks as internal partitioning tied into standard bricks see pages 34&35.

Bricklaying Practice *(Should conform to the current 'BCA' and the 'AS 3700 Masonry Code').*

Bricks with a blend or texture should be laid so that the blend is evenly distributed and that the texture is sloping down to prevent dirt collecting. Bricks should be laid with full mortar beds and perpends covering the full brick width. When laying, bricks should be cleaned off at regular intervals with a brush and also a moist sponge if necessary. Joints *should not* be greater than 10mm thick. Raked joints *should not* be deeper than 10mm and *should not* be used in saline (salt) or heavy industrial environments.

Termite Control in Brick Veneer

Brick veneer walls which have in the past been taken below ground level have provided termites with easy access to the structural frame. Hence the use of stainless steel mesh products to the perimeter and rebate. When the footing is raised above ground, the slab edge can be utilised as a termite deterrent, see Pages 15&16.

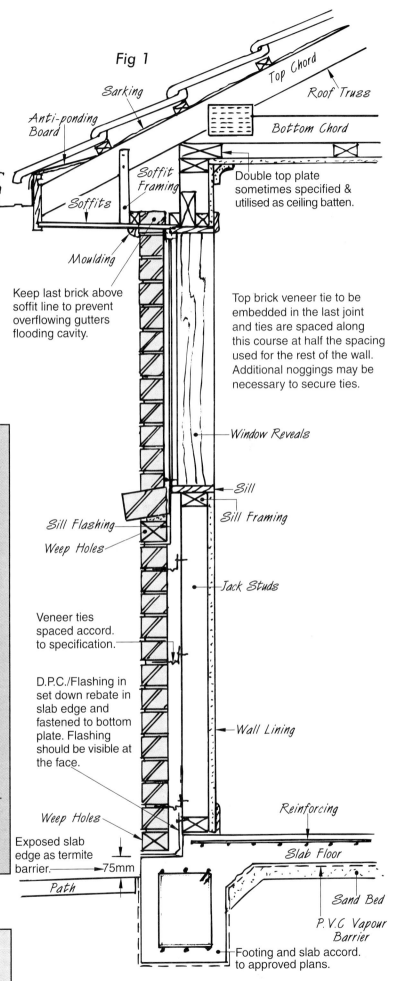

Fig 1

Sarking

Top Chord

Roof Truss

Anti-ponding Board

Bottom Chord

Soffit Framing

Double top plate sometimes specified & utilised as ceiling batten.

Soffits

Moulding

Keep last brick above soffit line to prevent overflowing gutters flooding cavity.

Top brick veneer tie to be embedded in the last joint and ties are spaced along this course at half the spacing used for the rest of the wall. Additional noggings may be necessary to secure ties.

Window Reveals

Sill

Sill Framing

Sill Flashing

Weep Holes

Jack Studs

Veneer ties spaced accord. to specification.

D.P.C./Flashing in set down rebate in slab edge and fastened to bottom plate. Flashing should be visible at the face.

Wall Lining

Reinforcing

Weep Holes

Exposed slab edge as termite barrier.

75mm

Path

Slab Floor

Sand Bed

P.V.C Vapour Barrier

Footing and slab accord. to approved plans.

BRICK VENEER SINGLE STOREY WITH STIFFENED RAFT SLAB FLOOR

Mortar Mixtures *(see also the A.S. on concrete masonry)*

a). M3 applications: Above DPC but subject to non-saline wetting and drying. Also site must be 100m away from a non-surf coast or 1km away from a surf coast:

1:1:6 - one part type GP or GB cement, one part hydrated lime and six parts bricklaying sand; or 1:0:5 plus the use of methyl cellulose water thickener.

b). M4 application: Below DPC in aggressive soils, or standing in salty or contaminated water as well as tidal or splash localities, or within 1km of an industry producing or dispersing chemical pollutants. Also suitable for above DPC within 100m of a non-surf coast or 1km of a surf coast:

1:½:4½ - one part type GP or GB cement, ½ part hydrated lime and 4½ parts bricklaying sand; or 1:0:4 plus the use of methyl cellulose water thickener.

Notes:

a). As very few bricklayers will be informed whether the sub-soil is aggressive or not, it will be prudent to use the M4 mixture for below DPC on all sites.

b). *Do not* use brickies loam.

Hint: If the brick veneer walls are to be built in two stages, it is advisable when ordering the base bricks to include the total brick requirement for the whole house. Variations in colour can occur when ordering separate lots.

2-Storey Brick Veneer

2-Storey construction, especially with cantilevered floors reduces the area of land use and can be more economical to build because of the reduction in footings and roof areas even though additional labour costs are incurred.

Joist Flooring System

It is essential that all floor framing be designed to effect a minimum of shrinkage. *See more on this subject on Page 57.*

Fig 1

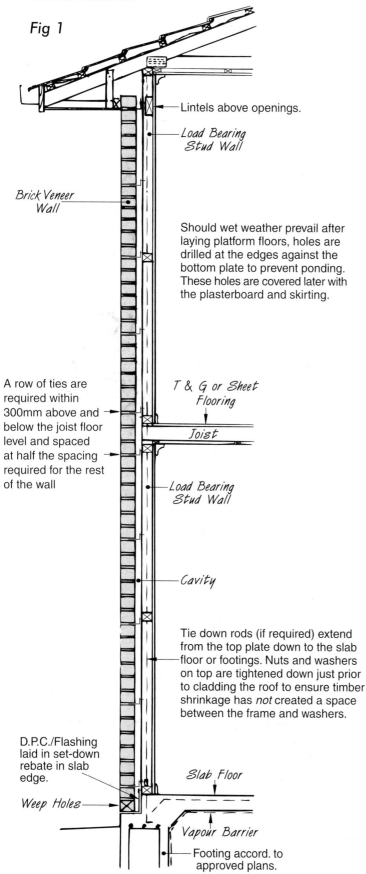

Lintels above openings.

Load Bearing Stud Wall

Brick Veneer Wall

Should wet weather prevail after laying platform floors, holes are drilled at the edges against the bottom plate to prevent ponding. These holes are covered later with the plasterboard and skirting.

A row of ties are required within 300mm above and below the joist floor level and spaced at half the spacing required for the rest of the wall

T & G or Sheet Flooring

Joist

Load Bearing Stud Wall

Cavity

Tie down rods (if required) extend from the top plate down to the slab floor or footings. Nuts and washers on top are tightened down just prior to cladding the roof to ensure timber shrinkage has *not* created a space between the frame and washers.

D.P.C./Flashing laid in set-down rebate in slab edge.

Slab Floor

Weep Holes

Vapour Barrier

Footing accord. to approved plans.

Fig 2

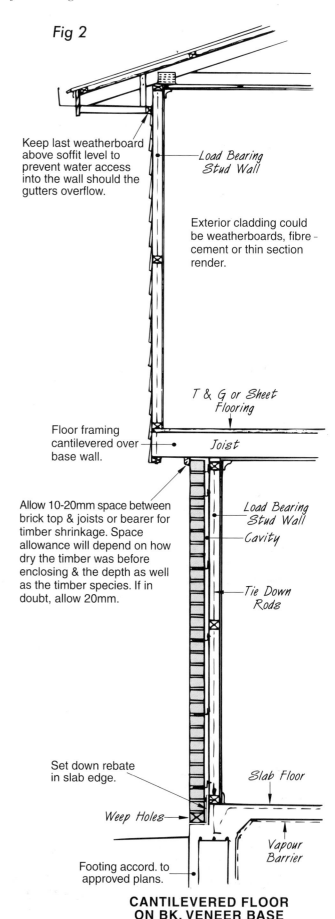

Keep last weatherboard above soffit level to prevent water access into the wall should the gutters overflow.

Load Bearing Stud Wall

Exterior cladding could be weatherboards, fibre-cement or thin section render.

T & G or Sheet Flooring

Joist

Floor framing cantilevered over base wall.

Allow 10-20mm space between brick top & joists or bearer for timber shrinkage. Space allowance will depend on how dry the timber was before enclosing & the depth as well as the timber species. If in doubt, allow 20mm.

Load Bearing Stud Wall

Cavity

Tie Down Rods

Set down rebate in slab edge.

Slab Floor

Weep Holes

Vapour Barrier

Footing accord. to approved plans.

Brick Veneer Footing & Slab Edge Details

(For 'Formwork', see Pages 47 & 48)

Slab Edge Set Down Rebate

A set down rebate in the edge of the slab floor around the perimeter is required. This prevents water crossing the cavity under the timber bottom plate during wet weather.

Cavity Width— The BCA permits a 25mm MIN. width between the brickwall and services or bracing. 40mm is normally provided between the brick wall and the stud frame. This includes the services.

Rebate Width— The rebate must be one full brick width plus the cavity width. This total is usually 150mm using standard 110mm wide bricks with a typical 40mm cavity width. Keep in mind that the brick veneer must *not* overhang the footing more than 15mm.

Rebate Depth — While the BCA permits this to be 20mm MIN., 35mm is a more appropriate minimum. However other factors must be considered including: *Minimum height above finished ground level, required by some lending institutions- which could be as much as 300mm; overflow relief gulleys and termite barriers*.

Overflow Relief Gulleys (ORG)

These must be 150mm MIN. above the lowest sanitary fitting. This is usually a floor waste but could be a sunken shower waste which is usually 50-75mm lower again.

Termite Barriers

Termite barriers must be installed in accordance with AS 3660.1 and the BCA. The slab floor can be an acceptable termite berrier if 75mm MIN. of the edge is exposed above ground or paving wither horizontally or vertically as in Fig 1 below or Fig 2 on Page 16 and the slab is engineer designed to prevent cracking and thus prevent termite access. Service penetrations through the slab must receive an additional termite barrier such as stainless steel mesh.

Exposing the slab edge enables visual inspection for termite mud tubes. Particular care should be paid to areas where services such as downpipes cross the exposed slab edge. These need to clear the concrete termite barrier for easy inspection.

This concrete termite barrier will necessitate the use of raft slabs as in fig 2 or stiffened raft slabs as in fig 1. Raft slabs incorporate the footing so that the concrete is poured to both in order to form a monolithic mass. Both designs (figs 1 & 2) will enable the use of polystyrene waffle pod bases should you choose *(see Page 52 for 'Waffle Pods')*.

Slab Edge Treatment

The exposed 75mm is *not* permitted to be rendered, clad, tiled or flashed. However, bagging, while *not* provided for in the BCA, could be applied using a thin oxide-tinted slurry to stain the edge in order to match bricks above. The slurry must be thin enough to only constitute a staining *not* a coating. The intent of the Codes ruling is that a coating could become loose thus providing access for termites behind the coating. Some states such as Qld permit a 100% acrylic paint system to be applied. Check with your local gov't building dept.

> **Note:** See pages 47 & 48 for 'How to Construct Formwork'.

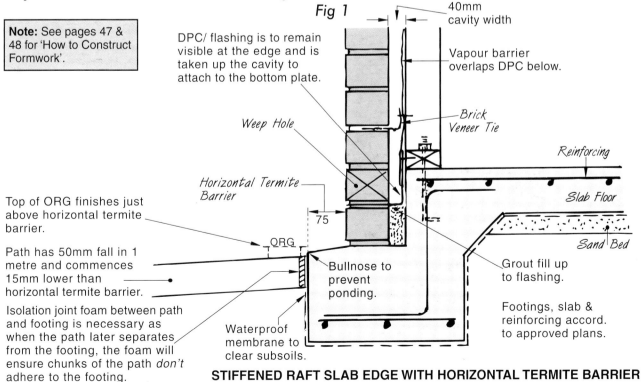

STIFFENED RAFT SLAB EDGE WITH HORIZONTAL TERMITE BARRIER

IMPORTANT NOTE: While the B.C.A permits the use of the slab & its edge to be designed as a termite barrier, it can only be effective if the edge is formed accurately, not requiring building out with render. On-site experience proves this does not reliably occur. Therefore an additional termite barrier should be provided either by using treated timber or cyprus thoughout or by S.S mesh etc. Chemicals as a barrier even when reticulation is installed can not be guaranteed as a permanent solution.

75mm Horizontal Slab Edge Termite Barriers *(fig 1, Page 15)*

This option is being adopted by many progressive builders for three reasons:

a). The vertical barrier requires the outer formwork to be straight without any deformity as rendering to straighten is *not* permitted. This is difficult to achieve hence the preference for the horizontal barrier.

b). The 75mm exposed strip *doesn't* require colouring to match adjacent bricks or finishes.

Other Considerations for the above

The 150mm MIN. height of the lowest sanitary fitting above the ORG and the 75mm MIN. above finished paving for DPC/flashing will necessitate a higher rebate step, usually 172mm (two standard bricks of 86mm including mortar). The ORG must finish clear of the barrier and any paths.

Ponding on the Barrier

This is *not* permitted. However, this is *not* usually a problem as long as a large bullnose is applied to the edge to prevent a lip or proudness of the edge. Paths should be 15mm below the barrier.

step need *not* be so high to attain the 150mm MIN. from the ORG to the lowest sanitary fitting unless required to be by others.

Preparing the Formwork

The outer forms on the vertical barrier must be straight without hollows as rendering to straighten is *not* permitted. To enable easy straightening of forms use the laminated planking timber available and coat with polyurethane or similar for easy cleaning.

Stripping the Formwork of both Slab Edge Termite Barriers

This must be performed on the same day and while there is still sufficient cement paste within the surface to enable sponging smooth.

Vibration

Vibration is necessary on all slabs and footings. However, where the slab edge forms a termite barrier as above, it is essential to remove all entrapped air along these areas and bring excess cement paste to the surface for sponging smooth.

75mm Vertical Slab Edge Termite Barriers *(fig 2)*

One advantage of the vertical barrier is that the rebate

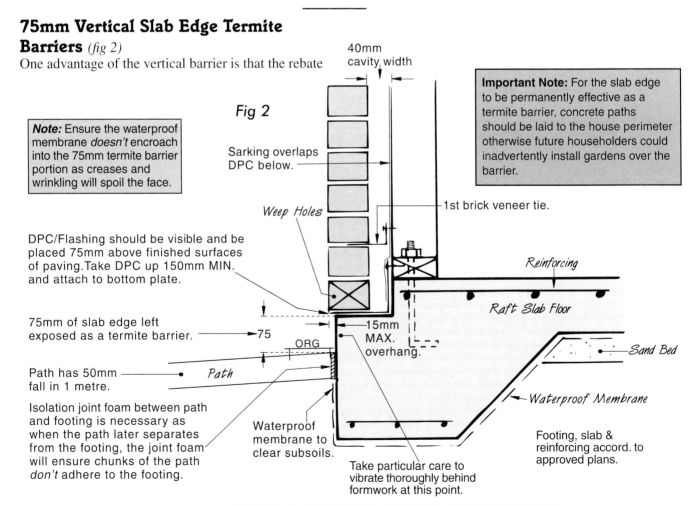

Fig 2

40mm cavity width

Note: Ensure the waterproof membrane *doesn't* encroach into the 75mm termite barrier portion as creases and wrinkling will spoil the face.

Sarking overlaps DPC below.

Important Note: For the slab edge to be permanently effective as a termite barrier, concrete paths should be laid to the house perimeter otherwise future householders could inadvertently install gardens over the barrier.

Weep Holes

1st brick veneer tie.

DPC/Flashing should be visible and be placed 75mm above finished surfaces of paving. Take DPC up 150mm MIN. and attach to bottom plate.

Reinforcing

Raft Slab Floor

75mm of slab edge left exposed as a termite barrier. ——75

ORG

15mm MAX. overhang.

Sand Bed

Path has 50mm fall in 1 metre.

Path

Isolation joint foam between path and footing is necessary as when the path later separates from the footing, the joint foam will ensure chunks of the path *don't* adhere to the footing.

Waterproof membrane to clear subsoils.

Waterproof Membrane

Take particular care to vibrate thoroughly behind formwork at this point.

Footing, slab & reinforcing accord. to approved plans.

RAFT SLAB EDGE WITH VERTICAL TERMITE BARRIER

Cont.

Brick Veneer on Sloping Sites

The design may utilise steps in the footings necessitating the brick walls to commence below ground. *Note:* this system is best to avoid if possible as it offers the greatest potential for termite access to the building. However, where it must be utilised, take care to have SS termite mesh applied by certified applicators. The footings are usually laid first then brick base walls contructed to slab height. If the step down in the rebate is one brick high, it will enable bricks to be temporarily utilised as formwork using weak mortar of 8:1 mix *(see fig 4, Page 48)*. The bricks can be later cleaned and reused. Alternatively, 145x45mm or 145x35mm timber could be fastened to the top of the brickwork extending into the proposed slab to form the rebate *(see figs 3 & 6, Page 48)*.

Bricks used Below Ground

Bricks laid below the damp proof course should be 'exposure grade quality' when used in aggressive soils or near the coast where M4 mortar is required *(see Page 13)*.

Damp Proof Course/Flashing

A damp proof course/flashing should be laid *not less than 150mm* above finished ground level or *not less than 75mm* above finished concrete paths or paving. *For other provisions see the BCA.* Embossed polyethylene sheet with an average thickness of 0.5mm is one DPC which may also be used as flashing. However embossed polyethylene coated aluminium will provide a longer life. Refer to AS/NZ 2904 for other approved DPC courses and flashings. On brick veneer a DPC doubling also as flashing is laid on the slab edge set-down as in figs 1 and 2,

Pages 15 & 16 and should extend to be visible at the outside face of the wall. When rendering brickwork or exterior cladding, this D.P.C. should extend through the render to remain visible.

Allow 150mm MIN. overlap at end-to-end joins. The D.P.C. is taken up the cavity 150mm MIN. above the bottom of the weep holes. It should cross the joint between the slab and timber bottom plate and is fastened to the framing (see fig 3).

Weep Holes

While the BCA permits weepholes to be spaced 1200mm MAX. apart, on standard metric bricks it is better to reduce this to every 4th perp. They should commence on the DPC/Flashing at the bottom of the cavity.

Paths

It is good practice to lay paths to the house perimeter. This helps to maintain the footing subsoils at a stable moisture content and prevent unequal subsoil movement. Paths should have a fall of 50mm in 1 metre into stormwater drainage which is piped to the street or approved disposal.

Location of Brick Walls on Footings

Brick walls should *not* overhang the footing or slab edge by more than 15mm (see fig 2, Page 16).

Where brick walls are located more centrally over footings below ground as in fig 3, the junction between footing and brickwork should be parged with mortar and the parging and brickwork up to just below paving surface given at least two coats of bituminous sealant such as Hydroseal or one thick coating of Hydroflex or similar.

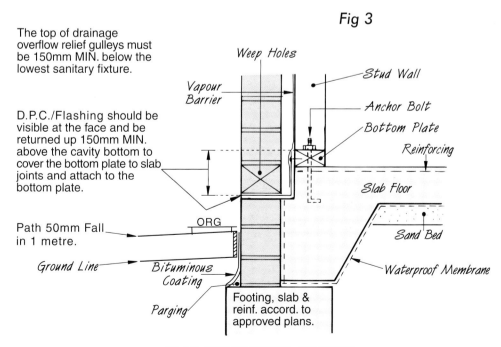

Fig 3

The top of drainage overflow relief gulleys must be 150mm MIN. below the lowest sanitary fixture.

D.P.C./Flashing should be visible at the face and be returned up 150mm MIN. above the cavity bottom to cover the bottom plate to slab joints and attach to the bottom plate.

Path 50mm Fall in 1 metre.

Ground Line

Bituminous Coating

Parging

Weep Holes

Vapour Barrier

ORG

Stud Wall

Anchor Bolt

Bottom Plate

Reinforcing

Slab Floor

Sand Bed

Waterproof Membrane

Footing, slab & reinf. accord. to approved plans.

SLAB SEPARATE TO FOOTING

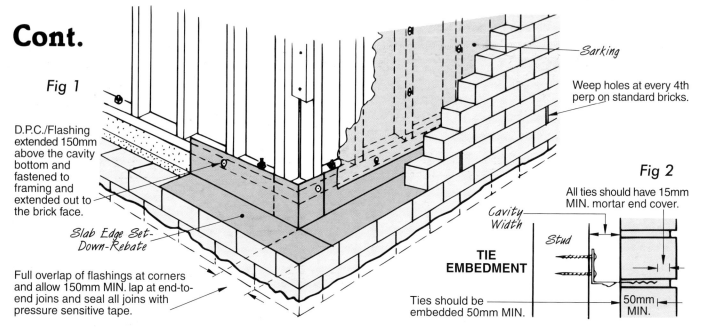

Fig 1

D.P.C./Flashing extended 150mm above the cavity bottom and fastened to framing and extended out to the brick face.

Slab Edge Set-Down-Rebate

Full overlap of flashings at corners and allow 150mm MIN. lap at end-to-end joins and seal all joins with pressure sensitive tape.

Sarking

Weep holes at every 4th perp on standard bricks.

Fig 2

All ties should have 15mm MIN. mortar end cover.

Cavity Width *Stud*

TIE EMBEDMENT

Ties should be embedded 50mm MIN.

50mm MIN.

Brick Veneer Ties

Approved brick veneer ties are attached to the timber frame at every stud and at vertical spacings as specified.

Tie packaging should have the category marked as well as their MAX. cavity width capability.

Selecting for Anti-Corrosion

It is essential to select the correctly rated ties for anti-corrosion.

R2 Steel ties with 300g/m² galvanising (z600) or better — (not suitable for aggressive soils) Suitable for all other areas except within 1km of a non-surf coast or 10km of a surf coast or below the D.P.C.

R3 Steel ties with 470g/m² galvanising or better Only suitable between 100m and 1km from a non-surf coast, between 1km and 10km from a surf coast and below the D.P.C in non-aggressive soils.

R4 — 316 Stainless steel. Required within 100m from a non-surf coast, within 1km from a surf coast, below the D.P.C in aggressive soils and in areas subject to saline wetting and drying.

For ties to give the builder long term protection from litigation, it is essential they be certified as tested.

Tie Placement — The first ties at the bottom should be in the first brick joint above the timber bottom plate and the last ties at the top should be embedded in the last joint.

On houses with timber floors ties should also be applied along the bearer 300mm apart at each floor level.

Also double the number of ties at the top of walls, at intersecting walls and around doors and window openings and articulation joints and immediately above and below an intermediate floor support. Where necessary, this can be achieved by fastening 2 ties at each stud either in adjacent bed joints or one tie on each side of the stud.

Ties should be horizontal or have a slight downward

slope away from the timber frame but the slope should *not* be greater than 10mm in the width of the cavity.

Fasteners: Use fasteners recommended by the manufacturer. These are required to be supplied by the manufacturer.

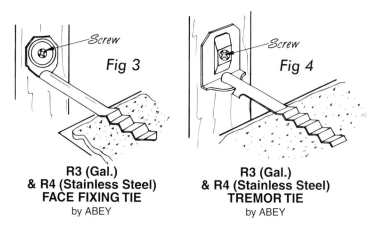

Screw

Fig 3

R3 (Gal.) & R4 (Stainless Steel) **FACE FIXING TIE** by ABEY

Screw

Fig 4

R3 (Gal.) & R4 (Stainless Steel) **TREMOR TIE** by ABEY

Tie Embedment: The tie is embedded in the brick joint 50mm MIN. as in fig 2.

Keeping Ties & Cavities Clean

This must be done to prevent water crossing the cavity and commencing decay in the structural frame.

When sarking is applied to the outside of the timber frame, cavities are cleaned by either suspending a batten down the cavity and raising it as work progresses or a brick is left out approx. every 1500mm along the cavity bottom and the mortar is hosed out regularly.

WALL TIE SPACINGS FOR BRICK VENEER For Wall Hts. 2.7m MAX.		
Wind Classification	**450 stud walls Horizontal x Vertical**	**600 stud walls Horizontal x Vertical**
N1—N4 & C1	450 x 600 M	600 x 600 M
C2 -- C3	450 x 600 M	600 x 600 H

Notes: a). 'M' = 'Medium Duty', 'H' = 'Heavy Duty', 'N' = 'Non Cyclonic', 'C' = 'Cyclonic' **b).** Double the number of ties beside control joints at the top of the wall, at intersecting walls and immediately above and below an intermediate floor support. Where necessart, this can be achieved by fastening two ties at each stud either in adjacent bed joints or one tie on each side of the stud.

Table used by kind permission from ABEY TIES. For Engineer related data contact ABEY AUST.

Brick Veneer Soffits & Joinery

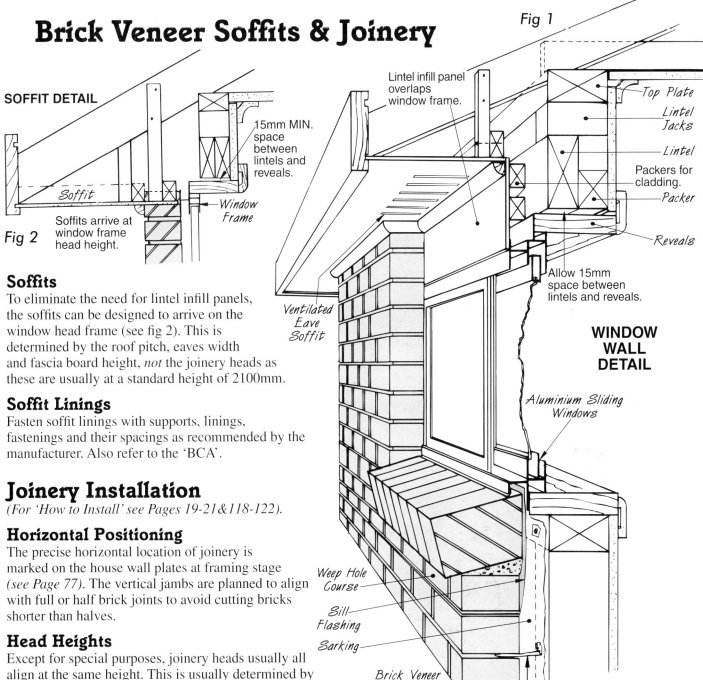

SOFFIT DETAIL

15mm MIN. space between lintels and reveals.

Soffit

Fig 2

Soffits arrive at window frame head height.

Window Frame

Fig 1

Lintel infill panel overlaps window frame.

Top Plate

Lintel Jacks

Lintel

Packers for cladding.

Packer

Reveals

Allow 15mm space between lintels and reveals.

WINDOW WALL DETAIL

Ventilated Eave Soffit

Aluminium Sliding Windows

Weep Hole Course

Sill Flashing

Sarking

Brick Veneer Ties

Soffits

To eliminate the need for lintel infill panels, the soffits can be designed to arrive on the window head frame (see fig 2). This is determined by the roof pitch, eaves width and fascia board height, *not* the joinery heads as these are usually at a standard height of 2100mm.

Soffit Linings

Fasten soffit linings with supports, linings, fastenings and their spacings as recommended by the manufacturer. Also refer to the 'BCA'.

Joinery Installation

(For 'How to Install' see Pages 19-21&118-122).

Horizontal Positioning

The precise horizontal location of joinery is marked on the house wall plates at framing stage *(see Page 77)*. The vertical jambs are planned to align with full or half brick joints to avoid cutting bricks shorter than halves.

Head Heights

Except for special purposes, joinery heads usually all align at the same height. This is usually determined by the height of standard aluminium patio doors above the floor which is commonly 2100mm. Where these are *not* used the standard entrance door height is used as the common height.

Waterproofing

The perimeter of joinery is a common location for moisture to gain access. Where the eaves provide

insufficient protection, flashings are installed above heads *(see fig 1, Page 20)*. Flashings to sills should be installed as in fig 3. These should have their ends enveloped as in figs 5-8, Page 20, to prevent water running straight through either end into the cavity. Side jamb flashings overlap the sill flashing as in fig 4. All flashings can be factory fitted. All joinery should be sealed around the perimeter with long life sealant after completion of brickwork.

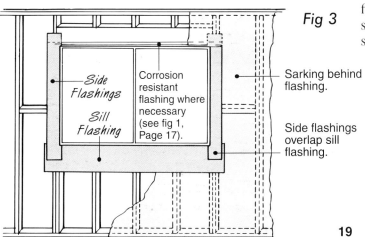

Fig 3

Side Flashings

Sill Flashing

Corrosion resistant flashing where necessary (see fig 1, Page 17).

Sarking behind flashing.

Side flashings overlap sill flashing.

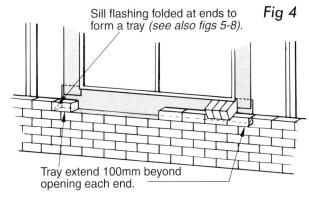

Sill flashing folded at ends to form a tray *(see also figs 5-8).*

Fig 4

Tray extend 100mm beyond opening each end.

Brick Veneer Joinery Cont.

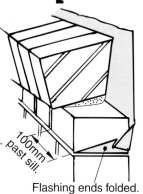

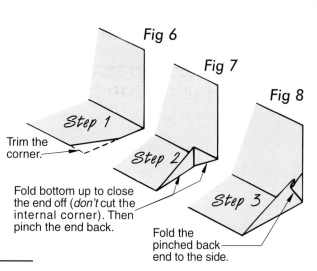

Fig 5

100mm past sill.

Flashing ends folded.

Fig 6

Fig 7

Fig 8

Step 1 — Trim the corner.

Step 2 — Fold bottom up to close the end off (*don't* cut the internal corner). Then pinch the end back.

Step 3 — Fold the pinched back end to the side.

Sill Flashing

Each end of sill flashings should be closed off by folding or enveloping into a tray shape. This is a simple exercise especially when using embossed plastic coated aluminium flashings or similar (see figs 6,7 & 8). This tray should extend to the next vertical brick joint or 100mm MIN. beyond the opening on both sides.

Binding (sticking) Joinery

On brick veneer it is not uncommon for windows or sliding patio doors to bind because of house frame shrinkage. To help avoid this the following space allowances should be provided at the following two points (as required in the 'BCA'). Double these clearances when using unseasoned hardwood (see also fig 2). Use an engineered joist system to reduce shrinkage to near zero.

a). 5mm between brick sills and window frames on the lower storey or single storey houses and 10mm on the second storey.

b). 8mm between top brick course and soffits on single storey houses and 12mm on the second storey.

Exterior Lintel Treatment

Lintels are often lined with infill panelling of fibre cement. The cladding overlaps the window frame as in fig 2 and its lower edge is embedded in a bead of sealant. Brickwork can be continued above openings as in fig 1. This is often required where bricks continue up into gable ends. M.S. angles are laid on the cavity side with the long leg placed vertically. These should be galvanised. Lintels should comply with the "R" ratings in the different exposure zones as for wall ties. Lintels should be stamped or labelled with their "R" rating.

Aluminium Windows in Brick Veneer

Windows are installed prior to laying bricks, so it is important to protect them from mortar and subsequent stains. Spray a protective coat of oil around frames and glass.

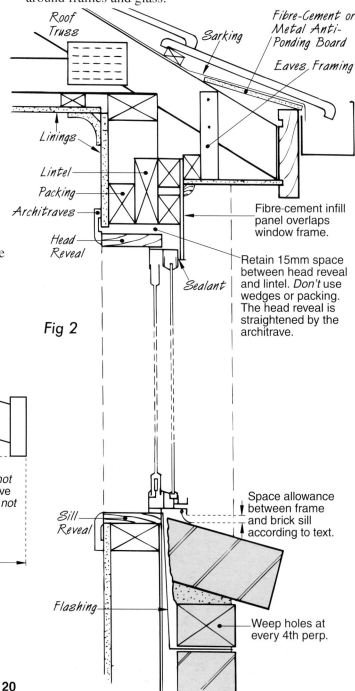

Roof Truss

Sarking

Fibre-Cement or Metal Anti-Ponding Board

Eaves Framing

Linings

Lintel

Packing

Architraves

Head Reveal

Fibre-cement infill panel overlaps window frame.

Retain 15mm space between head reveal and lintel. *Don't* use wedges or packing. The head reveal is straightened by the architrave.

Fig 2

Sealant

Sill Reveal

Flashing

Space allowance between frame and brick sill according to text.

Weep holes at every 4th perp.

Fig 1

Plaster Board

Lintel Flashing

Timber Lintel

Packing

15mm space

Reveals

Aluminium Frame

Glass

Sealant

M.S. Gal. Lintel Angle

Weep Holes

There should be a MIN. of three brick courses above M.S. lintels.

'H'

Lintel flashing is *not* required when eave depth as shown is *not* less than 3x'H'.

Eave Depth

Mild Steel Lintels

For M.S. Lintels sizes (see Table on Page 31).

Aluminium Doors in Brick Veneer with Slab Floor

(see also Page 119)

On concrete floors, doors should be installed into a set down rebate. *For timber floors, see Page 25.* The sill should be supported throughout its length and flashing should be laid underneath. Alternatively, it is better practice to install an aluminium sub-sill as seen in fig 3 and available from some suppliers. This sub-sill will provide a much better appearance to the front edge and also help prevent deflection of the sill which is a common occurrence. At brick lintels flashings are the same as in fig 1, Page 20. Jambs and flashings can be factory fitted.

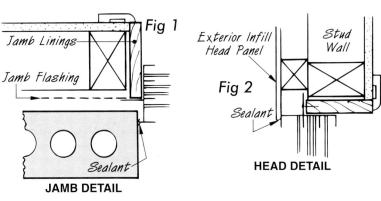

JAMB DETAIL

HEAD DETAIL

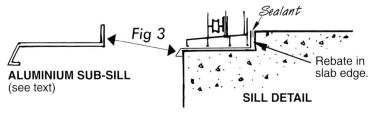

Fig 3
ALUMINIUM SUB-SILL
(see text)

SILL DETAIL

Timber Windows in Brick Veneer

Ensure all sides including backs and ends are thoroughly prime painted before installation. Provide sill flashings folded into a tray shape with ends folded up *(see Page 20)*.

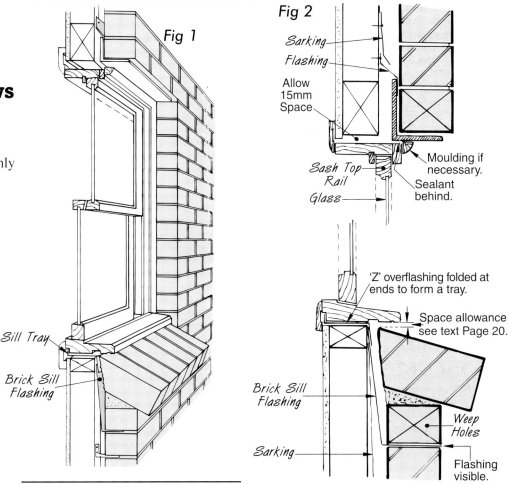

Exterior Timber Doors in Brick Veneer (Slab Floor)

All joinery should be either treated or hardwood of above ground durability Class 2 or better. These should be thoroughly prime painted on all sides and ends. Self closing door seals should be fitted to the threshold to resist driving rain.

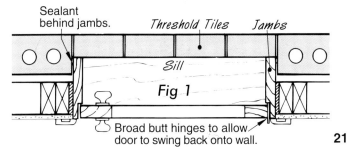

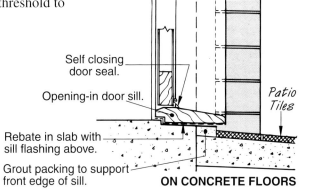

21

Brick Veneer with Brick Gables

Brick Gable with Roof Overhang

The brick walls can continue up into the gables. In figs 1 & 2 the roof is carried over the brick work but supported by the gable truss. Construction is as described on Page 94. *For 'cantilever & back span of members', see Page 94.*

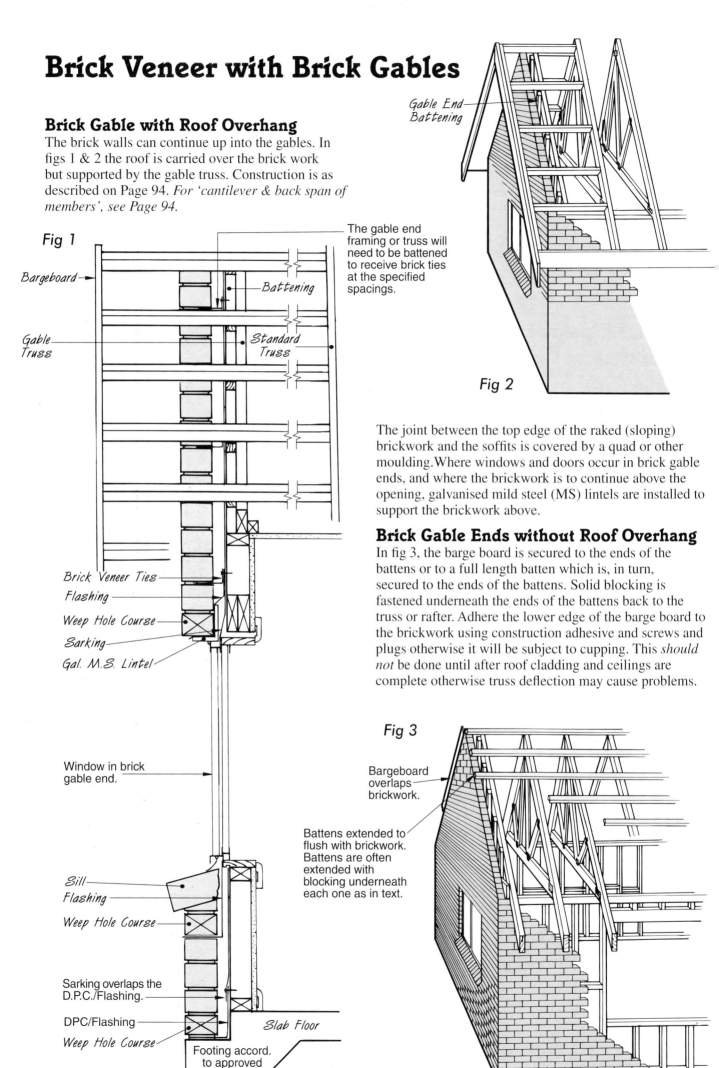

Gable End Battening

Fig 2

Fig 1

Bargeboard

Gable Truss

Battening

Standard Truss

The gable end framing or truss will need to be battened to receive brick ties at the specified spacings.

Brick Veneer Ties
Flashing
Weep Hole Course
Sarking
Gal. M.S. Lintel

Window in brick gable end.

Sill
Flashing
Weep Hole Course

Sarking overlaps the D.P.C./Flashing.

DPC/Flashing
Weep Hole Course
Slab Floor
Footing accord. to approved plans.

The joint between the top edge of the raked (sloping) brickwork and the soffits is covered by a quad or other moulding. Where windows and doors occur in brick gable ends, and where the brickwork is to continue above the opening, galvanised mild steel (MS) lintels are installed to support the brickwork above.

Brick Gable Ends without Roof Overhang

In fig 3, the barge board is secured to the ends of the battens or to a full length batten which is, in turn, secured to the ends of the battens. Solid blocking is fastened underneath the ends of the battens back to the truss or rafter. Adhere the lower edge of the barge board to the brickwork using construction adhesive and screws and plugs otherwise it will be subject to cupping. This *should not* be done until after roof cladding and ceilings are complete otherwise truss deflection may cause problems.

Fig 3

Bargeboard overlaps brickwork.

Battens extended to flush with brickwork. Battens are often extended with blocking underneath each one as in text.

Brick Veneer (with Timber Floor)

About Single Leaf Base Walls

Brick veneer walls with a timber floor have brick piers attached to the base of the walls at specified centres. Most of these piers will in turn support the floor bearers around the perimeter (see fig 1). The size and spacing of reinforced and unreinforced piers should be stipulated on approved plans. Footings are increased in width at pier locations as specified. Brick ties, flashings, weep holes, damp courses and other bricklaying requirements are the same as previously described for brick veneer with a slab floor.

Ant Capping (Shielding)

Continuous ant capping is built into the brickwork at the underside of the lowest timber framing members. *For correct application see Page 65.*

DPC

A DPC should be laid across the brickwork including engaged piers and *not* less than 150mm above finished ground level or paving and also directly below the bearer height on upper timber floors. *See Page 17 for suggested DPC types.*

Vermin Mesh (*see Page 25*)

Vermin must be prevented from entering the cavity by attaching fine wire netting to the bottom plate then across the cavity and embedded into the mortar joint. Any mortar droppings must be cleaned off the vermin proofing to prevent moisture crossing the cavity.

Ventilators (*see Page 25*)

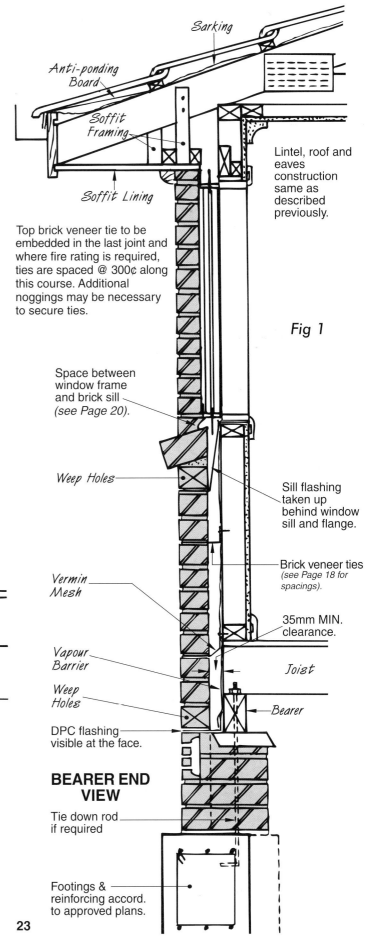

Sarking

Anti-ponding Board

Soffit Framing

Soffit Lining

Lintel, roof and eaves construction same as described previously.

Top brick veneer tie to be embedded in the last joint and where fire rating is required, ties are spaced @ 300¢ along this course. Additional noggings may be necessary to secure ties.

Fig 1

Space between window frame and brick sill (see Page 20).

Weep Holes

Sill flashing taken up behind window sill and flange.

Brick veneer ties (see Page 18 for spacings).

35mm MIN. clearance.

Vermin Mesh

Vapour Barrier

Joist

Weep Holes

Bearer

DPC flashing visible at the face.

BEARER END VIEW

Tie down rod if required

Footings & reinforcing accord. to approved plans.

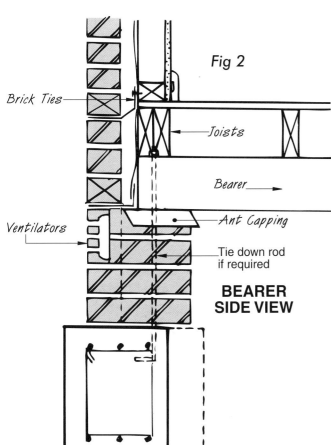

Fig 2

Brick Ties

Joists

Bearer

Ventilators

Ant Capping

Tie down rod if required

BEARER SIDE VIEW

Single Leaf Base Walls, Engaged & Isolated Piers

Brick veneer base walls and piers are constructed as in fig 1. Engaged piers may be unreinforced as in figs 2, 4 & 5 or reinforced as in fig 3, 6 & 7 according to approved plans.

Isolated Piers

These may also be unreinforced or reinforced according. to approved plans. Gal. steel columns are commonly used instead of brick.

Reinforced Isolated Brick Piers

Isolated piers supporting roof structures such as porches, verandahs and carports must be a MIN. of 350x350mm sq. x 2.7m MAX. in height and spaced at *not* greater than 3m centres with tie down and reinforcement as specified in approved plans and specifications. *See the 'AS 3700 Masonry Code' for MAX. heights and spacings when supporting various elements.*

Grout for Reinforced Piers

Grout should be 12MPa MIN. When mixing on-site, use 1 Portland Cement to 2 medium river sand to 4 of 10mm aggregate. Tamp well to remove entrapped air. *Above mix recommended by Cement, Concrete & Aggregates Australia.*

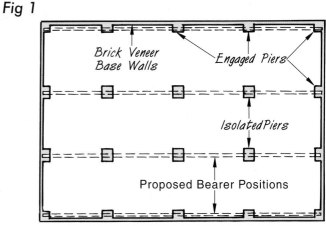

Fig 1

Brick Veneer Base Walls

Engaged Piers

Isolated Piers

Proposed Bearer Positions

BRICK VENEER BASE WALLS & PIERS

ENGAGED PIERS

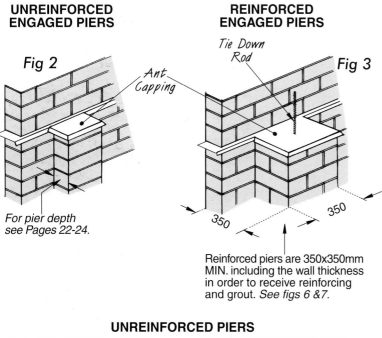

UNREINFORCED ENGAGED PIERS

Fig 2

Ant Capping

For pier depth see Pages 22-24.

REINFORCED ENGAGED PIERS

Tie Down Rod

Fig 3

350

350

Reinforced piers are 350x350mm MIN. including the wall thickness in order to receive reinforcing and grout. *See figs 6 & 7.*

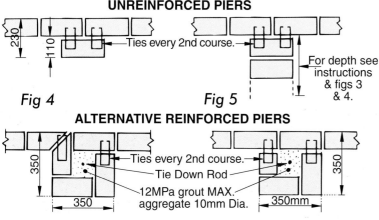

UNREINFORCED PIERS

230

110

Ties every 2nd course.

Fig 4

Fig 5

For depth see instructions & figs 3 & 4.

ALTERNATIVE REINFORCED PIERS

350

350

350

Ties every 2nd course.

Tie Down Rod

12MPa grout MAX. aggregate 10mm Dia.

350mm

Fig 6

Fig 7

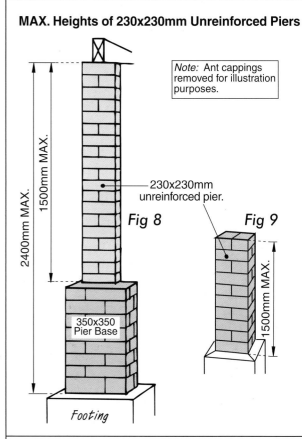

MAX. Heights of 230x230mm Unreinforced Piers

Note: Ant cappings removed for illustration purposes.

2400mm MAX.

1500mm MAX.

230x230mm unreinforced pier.

Fig 8

Fig 9

1500mm MAX.

350x350 Pier Base

Footing

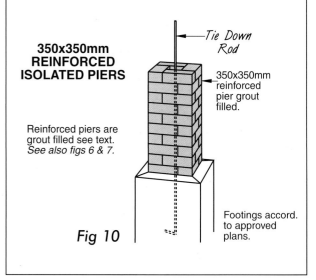

350x350mm REINFORCED ISOLATED PIERS

Tie Down Rod

350x350mm reinforced pier grout filled.

Reinforced piers are grout filled see text. *See also figs 6 & 7.*

Footings accord. to approved plans.

Fig 10

24

Ventilators

(Refer to the BCA for full details as the following is a guideline only). Approved ventilators with 7300mm^2 of clear opening per lineal metre of wall are positioned in walls immediately below bearer height and within 600mm of corners. Soldier bricks can also be used. Ventilators are vitally important as they help to keep subfloor timbers dry and so prevent decay commencing. *Avoid using ventilators which will corrode.* Ensure sufficient cross ventilation is provided to enclosed corners and spaces and that ventilation holes or slots are screened behind to prevent vermin access.

> *Note:* When screening is applied to ventilators, the net ventilation area of the vents is reduced by 50% so allow for this when selecting ventilation.

The Construction Procedure

The base brick walls and piers are built followed by the timber floor, walls and roof including installation of exterior joinery. Then the remaining brick walls are built.

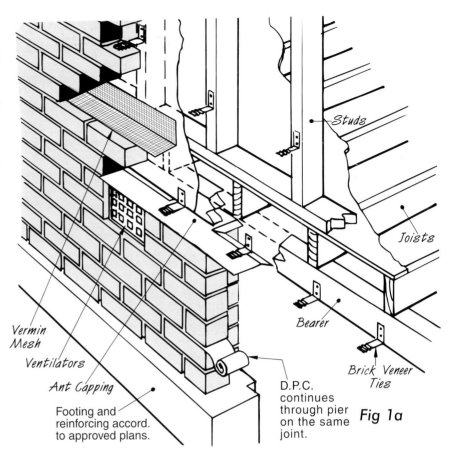

Vermin Mesh

Ventilators

Ant Capping

Footing and reinforcing accord. to approved plans.

Studs

Joists

Bearer

Brick Veneer Ties

D.P.C. continues through pier on the same joint.

Fig 1a

Installing Joinery in Brick Veneer
(with Timber Floor)

Installing Timber Door Frames

All frames should be treated or be of hardwood of above ground durability Class 2 or better. They should be thoroughly prime painted on all sides and ends before installation. Self closing door seals should be fitted to the threshold to resist driving rain.

Installing Aluminium & Timber Windows *Install the same as on Page 118.*

Installing Aluminium Sliding Doors

These can have timber jambs and flashings factory fitted. The aluminium sill is installed on the joists with flashing *(or an aluminium sub-sill, see Page 21)* underneath to protect the floor framing from moisture damage. The sill should be continuously supported with solid nogging or a full length of

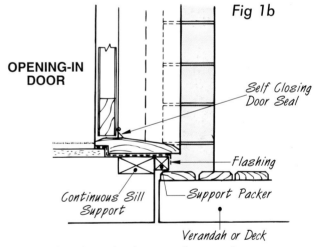

Fig 1b

OPENING-IN DOOR

Self Closing Door Seal

Flashing

Continuous Sill Support

Support Packer

Verandah or Deck

framing. The flooring edge is cut to fit up to the back of the sill with the flashing between. Flashings should be installed down the jambs in the cavity behind the door fins and in such a way as to prevent moisture gaining access to the house frame. *For 'how to install' see Page 119.*

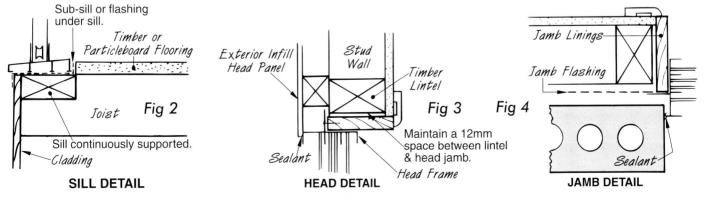

Sub-sill or flashing under sill.

Timber or Particleboard Flooring

Joist **Fig 2**

Sill continuously supported.

Cladding

SILL DETAIL

Exterior Infill Head Panel

Stud Wall

Timber Lintel

Fig 3

Maintain a 12mm space between lintel & head jamb.

Sealant

Head Frame

HEAD DETAIL

Jamb Linings

Jamb Flashing

Fig 4

Sealant

JAMB DETAIL

25

Cavity Brick Construction (Double Brick)

Cavity brick consists of two brick walls standing side-by-side with a cavity between and tied together with cavity ties. The inner brick leaf can be finished as face bricks, or common bricks can be used and walls rendered or lined with wallboard.

Internal partitions can be brick or concrete block and finished in a similar manner as above or timber frame partitions can be used and wallboard lining attached. Footings and slab should be designed by an Engineer.

On slab floors the outer brick leaf should be stepped down lower than the internal floor surface as seen in figs 1 & 2, Page 27. Ensure all brick joints are well filled, especially below ground level to prevent termite access.

Ceiling Binders omitted for illustration, see Page 102.

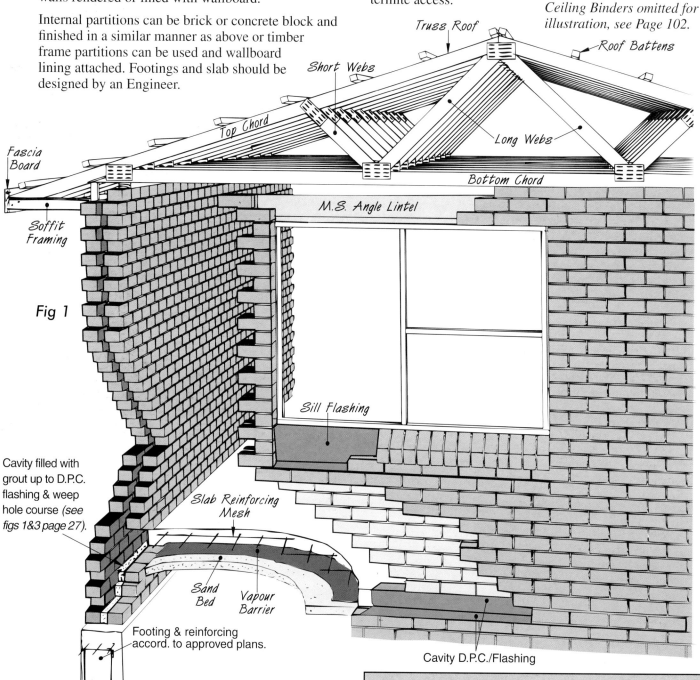

Truss Roof

Roof Battens

Short Webs

Top Chord

Long Webs

Fascia Board

Bottom Chord

Soffit Framing

M.S. Angle Lintel

Fig 1

Sill Flashing

Cavity filled with grout up to D.P.C. flashing & weep hole course *(see figs 1&3 page 27).*

Slab Reinforcing Mesh

Sand Bed

Vapour Barrier

Footing & reinforcing accord. to approved plans.

Cavity D.P.C./Flashing

Cavity Brick on Unstable Subsoils
On unstable subsoils cavity brick should be articulated. This means that vertical movement joints are provided at designated spacings on walls and between masonry and joinery to allow for footing movement to avoid or reduce cracking. *Consult the 'Cement, Concrete & Aggregates of Australia' in your State for literature on articulation.*

Stages of Construction
1. Footings are laid with reinforcing rods inserted and sanitary drainage installed.
2. Bricks are laid up to slab floor level or when using a timber floor, up to bottom of bearer height.
3. Floor is laid.
4. The inner leaf (skin) of the double wall is built with the windows and exterior doors built-in.
5. Internal partitions are constructed.
6. Roof is erected and clad.
7. Internal linings are attached and all interior doors fitted and finishing work carried out. At the same time the outer brick leaf on the double brick walls can be built.

Cavity Brick Footings & Slab Edge Details

The Set Down Rebate

A set down rebate in the edge of slab floors is required to prevent moisture crossing the cavity under the inner leaf. It is best to make this one full brick high to ensure the horizontal joints are on the same level as the inner leaf for embedding cavity ties. The width of the rebate needs to be one full brick width plus the cavity width. The cavity width *should not* be less than 35mm or wider than 65mm (50mm is a normal width) in cavity work. This clearance must be clear of bracing, services and insulation etc. One common practice is to pour the concrete footing and later form the slab edge using formwork *(see figs 7&8, Page 48)*. This method is used for fig 1 opposite.

Brickwork Below Ground Should be waterproofed and any footing-to-brick corners parged as in fig 1. Where fill is required under the slab as on sloping terrain, brickwork can be brought up from the footing as in fig 3. On raft slabs (see fig 2) the formwork is installed after excavating and the footings and slab concrete poured together *(see fig 2, Page 16)*. Formwork could be as in fig 1 or 2, Page 47. Where the footing is required to be extended above ground and the slab floor poured separately as in fig 9, Page 48, formwork is applied to both sides of the footing.

Termite Control with Slab Floors

While brick walls are *not* vulnerable to termite attack, the timber roof and ceiling framing are, so termite protection must be provided to conform to AS 3660.1. *(read Pages 15-17 also Page 141)*.

Slab Edge as a Perimeter Termite Barrier

When using the slab edge as a termite barrier, it must project above ground or paving 75mm MIN. vertically or horizontally and the exposed edge left smooth and free of air voids or ripples. Cement rendering or painting is *not* permitted as this can provide termite access through any cracks or drummy areas or behind loose paint. *For more details see Pages 15-17 also Page 141)*.

Brick Wall Overhanging Footings

Where footings extend above ground, the brickwork can be left flush or overhang the footing edge by 10mm but *shouldn't* exceed 15mm (see fig 2).

Weep Holes

While the 'BCA' permits weep holes to be spaced at 1200mm MAX. apart, it is good practice to reduce this to every 4th perp on standard bricks. The lower half of the perp should be cleaned out down to the DPC.

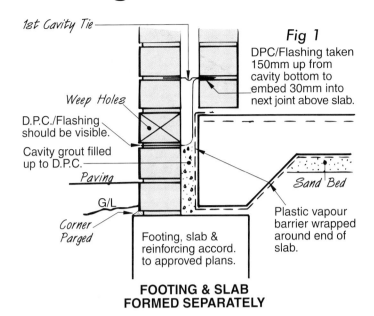

Fig 1

1st Cavity Tie

DPC/Flashing taken 150mm up from cavity bottom to embed 30mm into next joint above slab.

Weep Holes

D.P.C./Flashing should be visible.

Cavity grout filled up to D.P.C.

Paving

G/L

Corner Parged

Sand Bed

Plastic vapour barrier wrapped around end of slab.

Footing, slab & reinforcing accord. to approved plans.

FOOTING & SLAB FORMED SEPARATELY

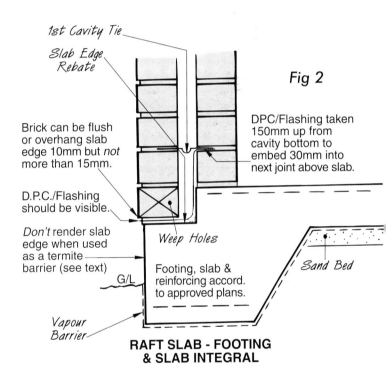

Fig 2

1st Cavity Tie

Slab Edge Rebate

Brick can be flush or overhang slab edge 10mm but *not* more than 15mm.

D.P.C./Flashing should be visible.

Don't render slab edge when used as a termite barrier (see text)

G/L

Weep Holes

Vapour Barrier

DPC/Flashing taken 150mm up from cavity bottom to embed 30mm into next joint above slab.

Sand Bed

Footing, slab & reinforcing accord. to approved plans.

RAFT SLAB - FOOTING & SLAB INTEGRAL

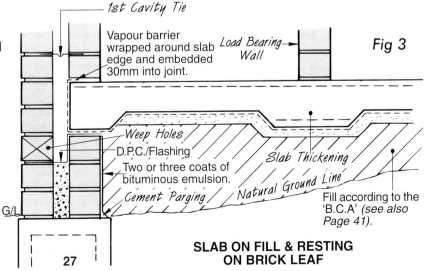

Fig 3

1st Cavity Tie

Vapour barrier wrapped around slab edge and embedded 30mm into joint.

Load Bearing Wall

Weep Holes

D.P.C./Flashing

Two or three coats of bituminous emulsion.

Cement Parging

Slab Thickening

Natural Ground Line

Fill according to the 'B.C.A' (see also Page 41).

G/L

SLAB ON FILL & RESTING ON BRICK LEAF

Cavity Brick Details

(see also AS/NZS 2904)

Damp Proof Courses

A damp proof course should be provided *not* less than 150mm above finished ground level or paving as well as directly below the bearer height on upper timber floors. On slab floors lay D.P.C. in the set-down as in fig 1. Bitumen coated aluminium 0.3-0.5mm thick is one approved DPC which may also be used as a flashing.

Cavity Flashings *(See figs 1 & 2, Page 27 for slab edge flashings).*

Where eaves provide insufficient weather protection to window heads, flashings should be provided above *(see figs 2, Page 31 and fig 7, Page 32)*. Provide weep holes every 4th perp. Weep holes are *not* required above or below openings less than 1m wide. Flashings should be provided below window sills *(see fig 2)*.

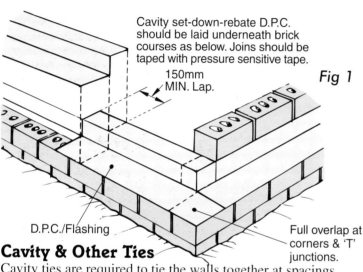

Cavity set-down-rebate D.P.C. should be laid underneath brick courses as below. Joins should be taped with pressure sensitive tape.

150mm MIN. Lap.

Fig 1

D.P.C./Flashing

Full overlap at corners & 'T' junctions.

Cavity & Other Ties

Cavity ties are required to tie the walls together at spacings according to the table below. If your locality has known seismic activity select ties to conform to AS 3700. Cavity ties are embedded across each brick leaf 50mm MIN. *For min. mortar end-cover see fig 1, Page 18.* Brick ties for cavity work are rated and colour coded as below. Cavity ties and the bottom of cavities should be kept free of mortar droppings. Use expansion ties across expansion joints.

Selecting for Anti-Corrosion

It is essential to select correctly rated ties for anti-corrosion. **R3 Steel ties with 470g/m² galvanising or better** only suitable between 100m and 1km from a non-surf coast, between 1km and 10km from a surf coast and below the D.P.C in non-aggressive soils.

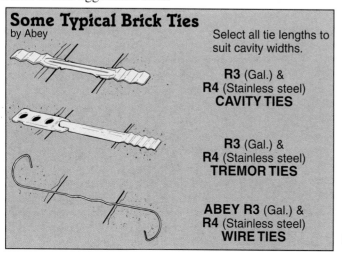

Some Typical Brick Ties
by Abey

Select all tie lengths to suit cavity widths.

R3 (Gal.) & **R4** (Stainless steel) **CAVITY TIES**

R3 (Gal.) & **R4** (Stainless steel) **TREMOR TIES**

ABEY R3 (Gal.) & **R4** (Stainless steel) **WIRE TIES**

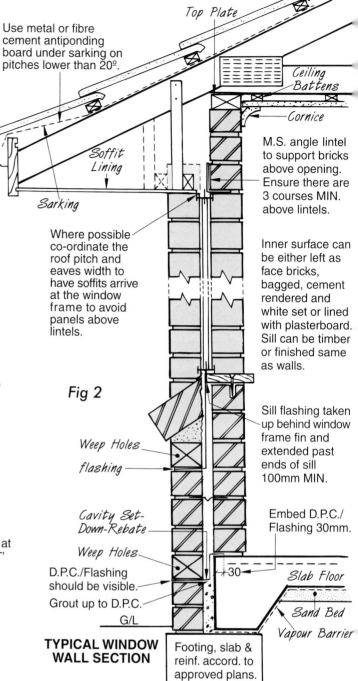

Use metal or fibre cement antiponding board under sarking on pitches lower than 20º.

Top Plate

Ceiling Battens

Cornice

Soffit Lining

Sarking

M.S. angle lintel to support bricks above opening. Ensure there are 3 courses MIN. above lintels.

Where possible co-ordinate the roof pitch and eaves width to have soffits arrive at the window frame to avoid panels above lintels.

Inner surface can be either left as face bricks, bagged, cement rendered and white set or lined with plasterboard. Sill can be timber or finished same as walls.

Fig 2

Weep Holes

flashing

Cavity Set-Down-Rebate

Weep Holes

D.P.C./Flashing should be visible. Grout up to D.P.C.

G/L

Sill flashing taken up behind window frame fin and extended past ends of sill 100mm MIN.

Embed D.P.C./ Flashing 30mm.

30

Slab Floor

Sand Bed

Vapour Barrier

TYPICAL WINDOW WALL SECTION

Footing, slab & reinf. accord. to approved plans.

R4 — 316 Stainless steel. Required within 100m from a non-surf coast, within 1km from a surf coast, below the D.P.C in aggressive soils and in areas subject to saline wetting and drying.

Wire Ties — All gal. wire ties must have 470g/M² Gal. and be coloured Red. For ties to give the builder long term protection from litigation, it is essential they be certified as 'Tested' and marked so on the packaging. Avoid ties without certification. Ties should be level or slope toward the outer leaf but *not* more than 10mm in the width of the cavity.

WALL TIE SPACINGS FOR CAVITY MASONRY						
Wind Classification	Horizontal (mm)	Vertical (mm)	Horizontal (mm)	Vertical (mm)	Horizontal (mm)	Vertical (mm)
N1 -- N2	300 x	600 M	450 x	600 M	600 x	600 M
N3	300 x	600 M	450 x	600 M	600 x	600 M
N4	300 x	600 M	450 x	600 M	600 x	600 M
C1 -- C2	300 x	600 M	450 x	600 M	600 x	600 H
C3	300 x	600 H	450 x	600 M	600 x	600 H

Notes: 'M' = 'Medium Duty', 'H' = 'Heavy Duty, 'N' = 'Non Cyclonic', 'C' = 'Cyclonic'

28 *Table used by kind permission from ABEY TIES. For Engineer related data contact ABEY AUST.*

Cont.

Joining Partitions into Cavity Walls

Fig 1

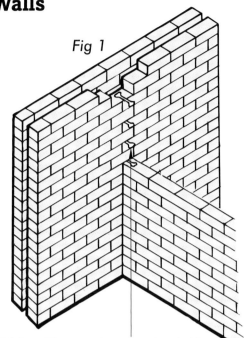

Brick partitions can be course bonded to the inner leaf of cavity walls or butt joined as above. In the latter case, ties are installed at alternate joints and embedded to a depth of 50mm MIN. into cavity wall and partition as above. The vertical joint must still be mortar bonded.

Fig 2

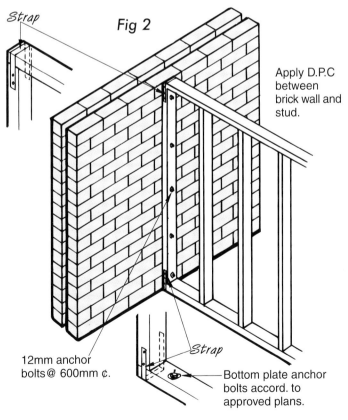

Apply D.P.C between brick wall and stud.

12mm anchor bolts @ 600mm ₵.

Strap

Bottom plate anchor bolts accord. to approved plans.

Bracing partitions are constructed as specified in approved plans and secured using 125x12mm masonry anchors at 600mm MAX. ₵. Bolts should penetrate brickwork 75mm MIN. and the top and bottom plates are strapped to the studs. Non-bracing timber partitions are secured to cavity walls through the end stud with masonry nails.

Roof Tie-Down

One nominal tie-down method approved for tiled or metal roofs in some N1 and N2 localities and where the roof span *does not* exceed 10m is depicted in fig 3. Figs 4 to 7 are typical methods applied in higher wind categories. Approved plans will specify the appropriate method.

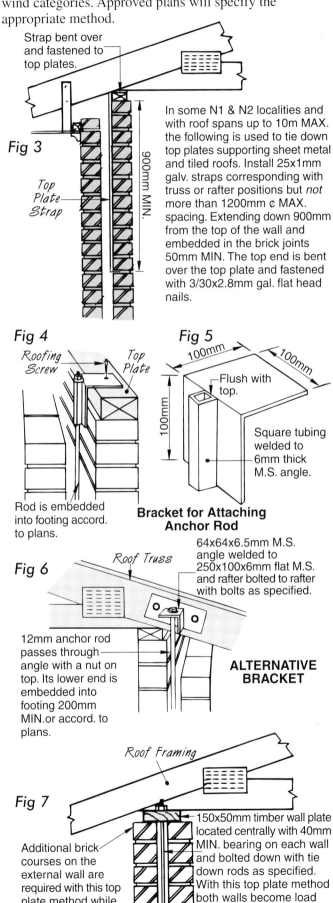

Strap bent over and fastened to top plates.

Fig 3

Top Plate Strap

900mm MIN.

In some N1 & N2 localities and with roof spans up to 10m MAX. the following is used to tie down top plates supporting sheet metal and tiled roofs. Install 25x1mm galv. straps corresponding with truss or rafter positions but *not* more than 1200mm ₵ MAX. spacing. Extending down 900mm from the top of the wall and embedded in the brick joints 50mm MIN. The top end is bent over the top plate and fastened with 3/30x2.8mm gal. flat head nails.

Fig 4

Roofing Screw

Top Plate

Rod is embedded into footing accord. to plans.

Fig 5

100mm

100mm

100mm

Flush with top.

Square tubing welded to 6mm thick M.S. angle.

Bracket for Attaching Anchor Rod

Fig 6

Roof Truss

64x64x6.5mm M.S. angle welded to 250x100x6mm flat M.S. and rafter bolted to rafter with bolts as specified.

12mm anchor rod passes through angle with a nut on top. Its lower end is embedded into footing 200mm MIN. or accord. to plans.

ALTERNATIVE BRACKET

Fig 7

Roof Framing

Additional brick courses on the external wall are required with this top plate method while in Figs 3 & 6 only the inner wall is load bearing.

150x50mm timber wall plate located centrally with 40mm MIN. bearing on each wall and bolted down with tie down rods as specified. With this top plate method both walls become load bearing. Rod is embedded 200mm MIN. into footing. or accord. to approved plans.

Cavity Brick & Timber Floor

The timber floor can be of platform or cut-in construction with timber partitions or when using brick partitions, the floor is cut in between the walls. Keep all timber framing and flooring 12mm clear of brick walls to allow for expansion of the timber.

Construction

The floor bearers are laid on engaged or attached brick piers and isolated piers, or columns of brick, concrete or steel are constructed for internal spans. Some situations will require the inclusion of tie down rods. This will necessitate brick piers to be increased from 230mm square to 350mm square to provide space for the rod and grout as in fig 10, Page 24.

Termite Control

Ant capping should be applied to all piers and base walls (see Page 65).

Ventilators

Approved ventilators must be built into walls as in text on Page 25. On cavity brick, an opening is provided directly behind the ventilators in the inner leaf (see figs 1&2).

Engaged Attached Piers

For plan view and construction design details, see Page 24.

Piers Without Tie-Down

Where tie-down is *not* required, piers supporting bearers could be reduced to one brick wide dependant upon the engineer's approval. These are tied to the inner leaf of the cavity wall with ties at alternate joints.

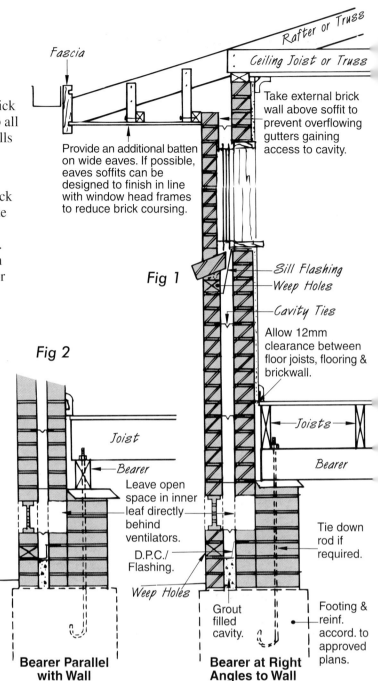

Fig 1

Fascia

Rafter or Truss

Ceiling Joist or Truss

Provide an additional batten on wide eaves. If possible, eaves soffits can be designed to finish in line with window head frames to reduce brick coursing.

Take external brick wall above soffit to prevent overflowing gutters gaining access to cavity.

Sill Flashing
Weep Holes
Cavity Ties

Allow 12mm clearance between floor joists, flooring & brickwall.

Joists

Bearer

Fig 2

Joist

Bearer

Leave open space in inner leaf directly behind ventilators.

D.P.C./ Flashing.

Weep Holes

Grout filled cavity.

Tie down rod if required.

Footing & reinf. accord. to approved plans.

Bearer Parallel with Wall

Bearer at Right Angles to Wall

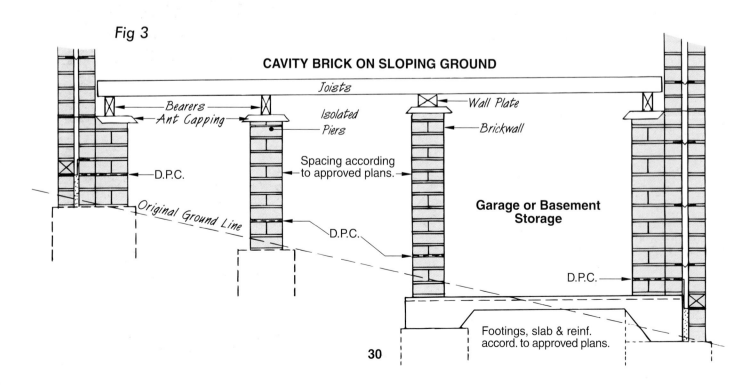

Fig 3

CAVITY BRICK ON SLOPING GROUND

Joists

Bearers
Ant Capping

Isolated Piers

Wall Plate

Brickwall

Spacing according to approved plans.

D.P.C.

Original Ground Line

D.P.C.

Garage or Basement Storage

D.P.C.

Footings, slab & reinf. accord. to approved plans.

Installing Windows into Cavity Brick

ALUMINIUM WINDOWS

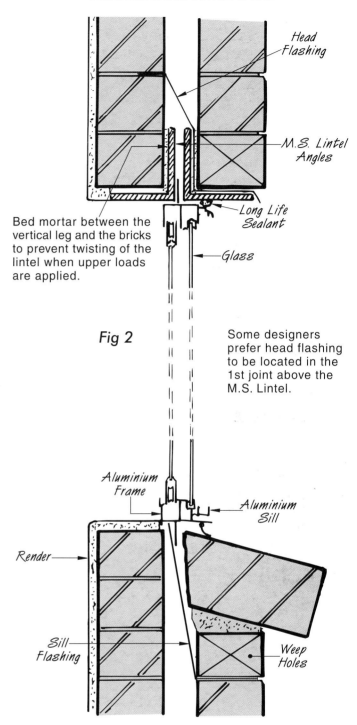

Windows are installed when the inner brick leaf arrives at sill height and are tied into brick reveal joints (see fig 3). All unreinforced brickwork above openings must be supported by lintels or arches. Gal. mild steel arch bars or angles are installed in each brick leaf above windows (see table below). These should have their ends bearing 100mm MIN. on each side of the opening. The long side of unequal angles is placed vertically. Ensure that any lintel loads will *not* bear onto the window frame.

Head flashings are necessary when eaves provide insufficient weather protection. Flashings which corrode can *not* be replaced, so ensure non-corrosive ones are used. Spray a protective coating of RP7 on aluminium frames and glass before cement droppings come into contact to prevent cement staining.

Bed mortar between the vertical leg and the bricks to prevent twisting of the lintel when upper loads are applied.

Fig 2

Some designers prefer head flashing to be located in the 1st joint above the M.S. Lintel.

GAL. MS LINTELS

MS LINTELS for 110 & 90mm SINGLE STOREY Supporting Loads (as in fig 1 - Type 'D')	
Span (mm)	M.S. Angle (mm)
1570	90x90x6EA
1810	100x100x6EA
1930	100x100x8EA
2770	150x90x8UA
3010	150x100x10UA

Span Lengths and angle sizes taken from Amend. 3 of the 'BCA' by kind permission of the 'Australian Building Codes' board.

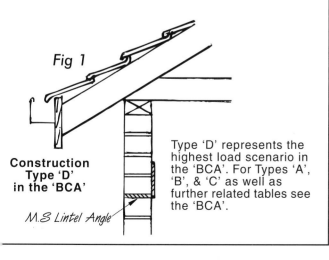

Fig 1

Construction Type 'D' in the 'BCA'

M.S Lintel Angle

Type 'D' represents the highest load scenario in the 'BCA'. For Types 'A', 'B', & 'C' as well as further related tables see the 'BCA'.

Fig 3
PLAN VIEW

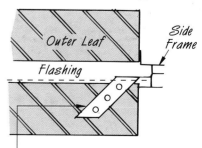

Building-in-lug every third course in N1, 2 & 3 wind categories.

TIMBER WINDOWS

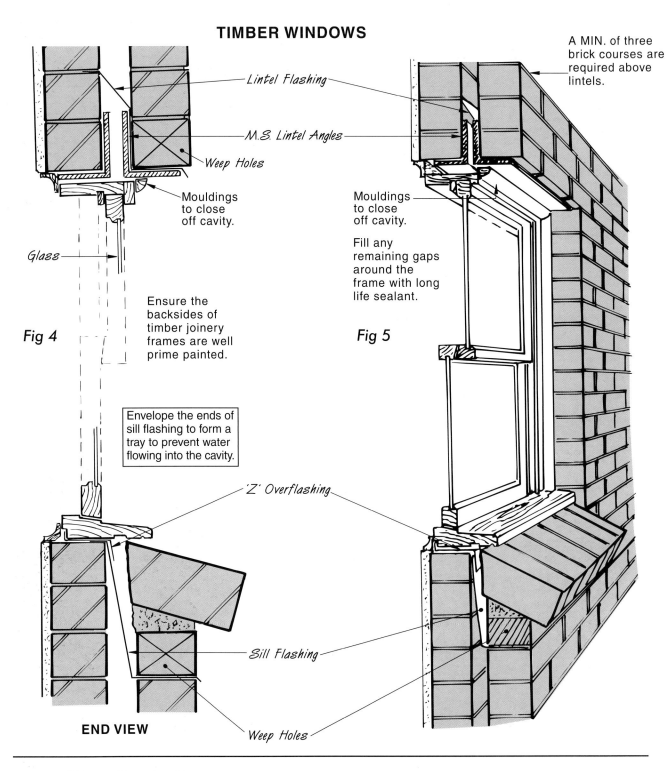

A MIN. of three brick courses are required above lintels.

Lintel Flashing

M.S. Lintel Angles

Weep Holes

Mouldings to close off cavity.

Mouldings to close off cavity.

Fill any remaining gaps around the frame with long life sealant.

Glass

Ensure the backsides of timber joinery frames are well prime painted.

Fig 4

Fig 5

Envelope the ends of sill flashing to form a tray to prevent water flowing into the cavity.

'Z' Overflashing

Sill Flashing

Weep Holes

END VIEW

Weep Holes

Installing Patio Doors

On concrete floors doors are installed into a rebate as in fig 6 and the sill should be continuously supported. Sill installation into timber floors is the same as in fig 1, Page 25. If the frame isn't wide enough to seal off the cavity, an aluminium cavity closer should be attached to the frame. Seal the frame perimeter with long life sealant.

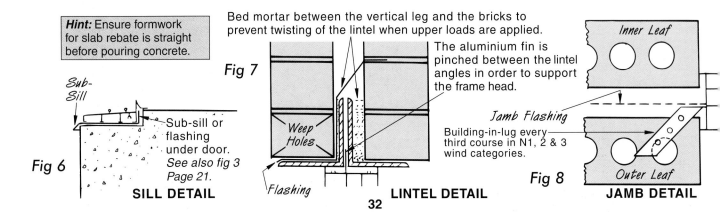

Hint: Ensure formwork for slab rebate is straight before pouring concrete.

Bed mortar between the vertical leg and the bricks to prevent twisting of the lintel when upper loads are applied.

The aluminium fin is pinched between the lintel angles in order to support the frame head.

Sub-Sill

Sub-sill or flashing under door. See also fig 3 Page 21.

Fig 6

Fig 7

Weep Holes

Flashing

SILL DETAIL

LINTEL DETAIL

Inner Leaf

Jamb Flashing

Building-in-lug every third course in N1, 2 & 3 wind categories.

Fig 8

Outer Leaf

JAMB DETAIL

Cavity Brick Details

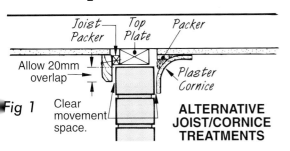

Fig 1

Joist Packer · Top Plate · Packer

Allow 20mm overlap

Clear movement space.

Plaster Cornice

ALTERNATIVE JOIST/CORNICE TREATMENTS

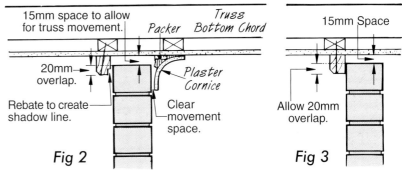

15mm space to allow for truss movement.

Packer · Truss Bottom Chord

20mm overlap.

Rebate to create shadow line.

Plaster Cornice

Clear movement space.

Fig 2

15mm Space

Allow 20mm overlap.

Fig 3

ALTERNATIVE TRUSS BOTTOM CHORD/CORNICE TREATMENTS

Brick Partition — Ceiling Joist Details

Allowance should be provided along the lower edge of the cornice line for movement (see fig 1). On truss roofs provide 15mm space between top of brick wall and ceiling lining for truss deflection. *Figs 2 & 3 provide alternative cornice treatments.*

Installing Door Frames into Cavity Brick Walls

Metal door frames are ideal for installing into brick walls. They are plumbed and temporarily supported by props with nails protruding down the top ends. These hook into the hollow frame. The props are secured with loose bricks as in fig 3. Ensure the hollow section is properly filled with mortar.

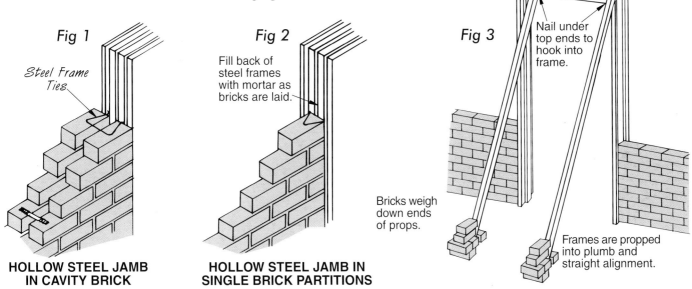

Fig 1

Steel Frame Ties

Fig 2

Fill back of steel frames with mortar as bricks are laid.

Fig 3

Nail under top ends to hook into frame.

Bricks weigh down ends of props.

Frames are propped into plumb and straight alignment.

HOLLOW STEEL JAMB IN CAVITY BRICK

HOLLOW STEEL JAMB IN SINGLE BRICK PARTITIONS

Installing Exterior Timber Door Frames into Cavity Brick

Ensure the back of frames and all end wood grains are well prime painted. Seal joint between frames and brick work with polyurethane sealant.

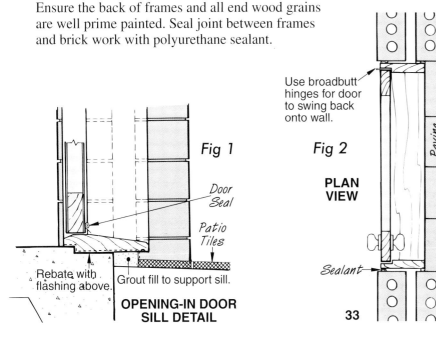

Fig 1

Door Seal

Patio Tiles

Rebate with flashing above.

Grout fill to support sill.

OPENING-IN DOOR SILL DETAIL

Use broadbutt hinges for door to swing back onto wall.

Fig 2

PLAN VIEW

Paving

Sealant

Wood jambs are fixed to each brick leaf with 32x0.8x300mm long galv. steel kinked straps at *not* less than 150mm or more than 400mm apart.

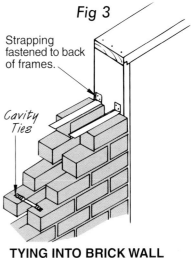

Fig 3

Strapping fastened to back of frames.

Cavity Ties

TYING INTO BRICK WALL

Cavity Brick with Dissimilar Internal Leaves or Walls

Cavity brick can be constructed with dissimilar or alternative internal leaves and walls.

Those commonly applied are:

Clay Presto, Pavilion or similar double height bricks, 90-150mm thick concrete block or AAC block, in which case 150mm thick would be used in the load bearing walls and 100mm in the non load bearing. For double height brick modules, see Page 12. *For all the above follow manufacturers specifications required for the situation.*

This system may enable faster and cheaper construction of the interior leaves and walls.

D.P.C and Cavity Ties

All walls are built on a D.P.C. slip joint at slab level (see fig 2). The BCA permits a 35-65mm MAX. cavity width between the two walls clear of pipes and hardware. However it is more customary to allow between 40-50mm. These walls are tied together with cavity ties at centres specified in approved plans and specifications. *See also 'Cavity Ties' table, Page 28.*

Placement of brick tie courses should be planned vertically to allow for the differing course heights to ensure the ties are laid horizontally or with only a slight outward slope.

> *Note:* It is important that all cavity ties are kept clean of any mortar or adhesive droppings which could create a bridge for moisture to cross to the inner load bearing wall.

Slab Edge Weepholes

The outer clay brick wall is built in a set down rebate at the slab edge. This should be a minimum of one brick course below the slab surface and weep holes are provided at every fourth perp.

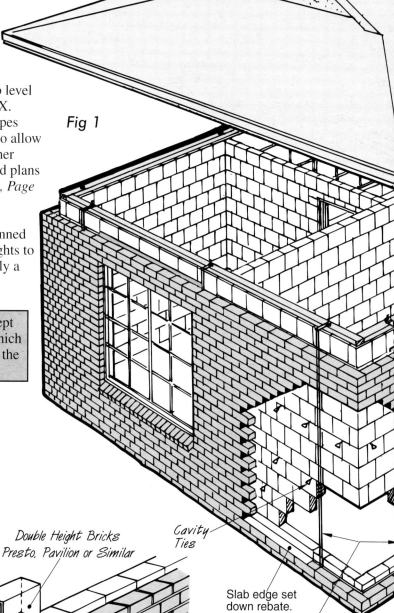

Fig 1

D.P.C. can be folded and laid in one piece or overlapped in two pieces, as illustrated. Fully overlap corners and provide a 150mm MIN. overlap at joins. Seal all joins with pressure sensitive tape. The first brick course above the D.P.C. is the weep hole course. Keep the cavity clear of mortar droppings.

PART 2

Folded D.P.C. in two parts.

Double Height Bricks Presto, Pavilion or Similar

Cavity Ties

PART 1

Slab edge set down rebate.

100mm MIN. overlap.

Weepholes

Fig 2

Tie down rods are taken down and embedded into footings or slab 200mm MIN.

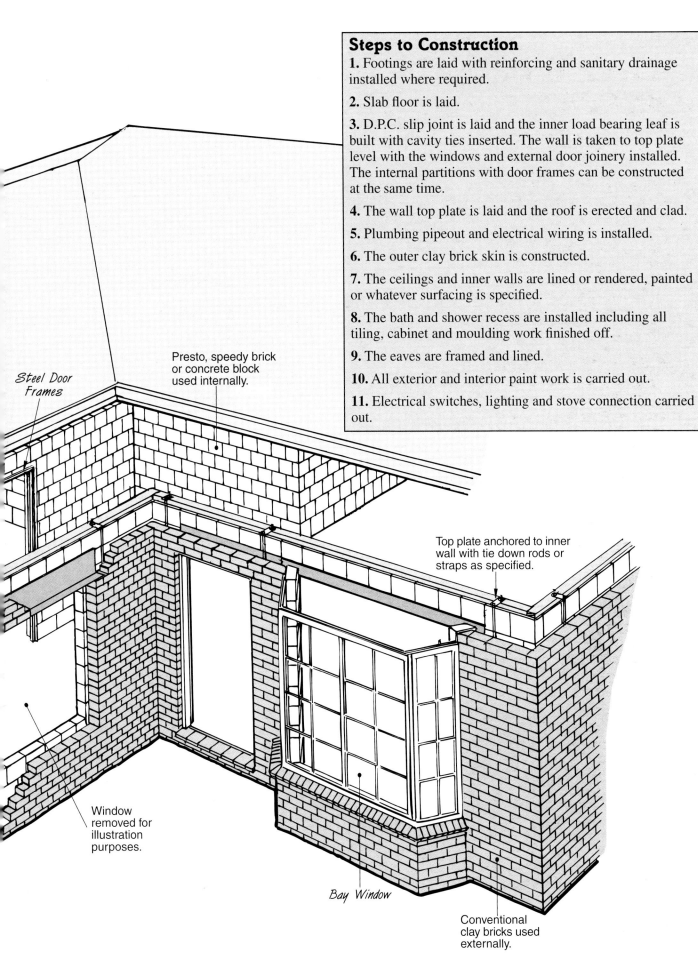

Steel Door Frames

Presto, speedy brick or concrete block used internally.

Top plate anchored to inner wall with tie down rods or straps as specified.

Window removed for illustration purposes.

Bay Window

Conventional clay bricks used externally.

Steps to Construction

1. Footings are laid with reinforcing and sanitary drainage installed where required.

2. Slab floor is laid.

3. D.P.C. slip joint is laid and the inner load bearing leaf is built with cavity ties inserted. The wall is taken to top plate level with the windows and external door joinery installed. The internal partitions with door frames can be constructed at the same time.

4. The wall top plate is laid and the roof is erected and clad.

5. Plumbing pipeout and electrical wiring is installed.

6. The outer clay brick skin is constructed.

7. The ceilings and inner walls are lined or rendered, painted or whatever surfacing is specified.

8. The bath and shower recess are installed including all tiling, cabinet and moulding work finished off.

9. The eaves are framed and lined.

10. All exterior and interior paint work is carried out.

11. Electrical switches, lighting and stove connection carried out.

Lintels

Mild steel lintel angles are required to support brickwork above openings. These should be hot dip galvanised. Install MS angles with the long side vertical *(see the Table on Page 31)*.

35

Concrete Block Construction (190mm commonly referred to 200mm thick single leaf) *For comprehensive local details and instructions see block manufacturer's instructions.*

190mm thick block walls used as the sole external walls are referred to as single leaf masonry. Special attention must be paid to waterproofing *(see Page 38)*. Linings can be fixed to blockwork internally. Partitions constructed in timber or 90 or 140mm blocks. Where reinforcing is required in internal walls 140 or 190mm blocks must be used. Where timber bracing partitions are used, they are bolted to block walls at 600mm centres and the end stud strapped to its top and bottom plates *(see fig 2, Page 29)*. Other partitions may be

hand or Ramset nailed. Co-operation will be required by the Plumber, Electrician and Carpenter whilst block work is in progress for the installation of services through and down the block cores.

Having Plans Drawn All block walls and openings should be block module length and height *(see Page 37)*. Ensure that size and positions of all reinforcing, control joints and bond beams are indicated clearly as well as the method of waterproofing.

Stages of Construction

1. Footings.
2. Any blockwork below slab level is laid.
3. Slab is prepared with all relevant drainage and plumbing services installed prior to pouring.
4. If using steel door frames, these are set up into plumb and straight alignment.
5. Blockwork is constructed with reinforcing, anchor bolts, plumbing and electrical conduits installed.
6. Concrete grout is poured to applicable vertical cores, lintels and bond beams.
7. Fix top plates to block walls if required and

any internal timber partitions constructed, then construct roof framing and eaves.

8. Attach sarking and cladding to roof.
9. Render exterior walls if required.
10. Paint walls throughout.
11. Install windows and exterior timber doors.
12. Batten and line ceilings. Line or render internal walls if required. Install bath and shower.
13. Install internal doors, kitchen cupboards, mouldings, ceramic tiling etc.

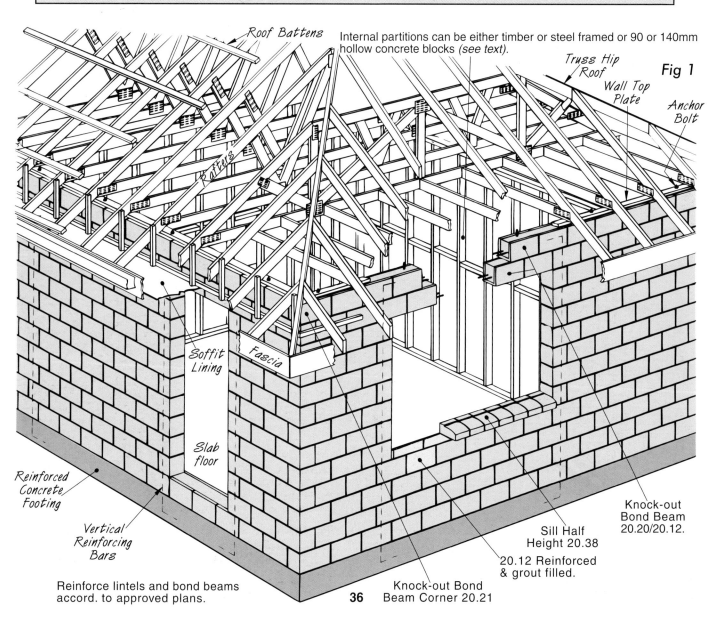

Roof Battens

Internal partitions can be either timber or steel framed or 90 or 140mm hollow concrete blocks *(see text).*

Truss Hip Roof

Fig 1

Wall Top Plate

Anchor Bolt

Rafters

Soffit Lining

Fascia

Slab floor

Reinforced Concrete Footing

Vertical Reinforcing Bars

Reinforce lintels and bond beams accord. to approved plans.

Knock-out Bond Beam Corner 20.21

Sill Half Height 20.38

20.12 Reinforced & grout filled.

Knock-out Bond Beam 20.20/20.12.

Blocklaying & Grouting

Blocklaying

Blocks must be kept dry on-site. At the end of each day's work, cover walls to prevent moisture entering the block cores. Blocks should be laid with the thicker part of the shell uppermost.

Joints Horizontal and vertical mortar joints should be 10mm thick and should be filled. Joints are ironed or filled flush *never* raked. Mortar overspread is removed and the blocks brushed clean. Any hard lumps or overspread can be removed by rubbing with a piece of block.

Note: Block walls are 'green' when freshly laid and at

Mortar Mixtures
(see also the AS 3700 AS4773.1 & AS4773.2 on concrete masonry)

a). M3 applications: Above DPC but subject to non-saline wetting and drying. Also site must be 100m away from a non-surf coast or 1km away from a surf coast:

> **1:1:6 — one part type GP cement, one part hydrated lime and six parts block laying sand; or 1:0:5 with methyl cellulose water thickener.**

b). M4 application: Below DPC in aggressive soils, or standing in salty or contaminated water as well as tidal or splash localities, or within 1km of an industry producing or dispersing chemical pollutants. Also suitable for above DPC within 100m of a non-surf coast or 1km of a surf coast:

> **1:½:4½ — one part type GP cement, ½ part hydrated lime and 4½ parts block laying sand; or 1:0:4 with methyl cellulose water thickener**

Notes: **a).** Should a water thickener be used, it must be Methyl Cellulose rather than plasticisers. Plasticisers can be detrimental to the mortar.

b). As very few blocklayers will be informed whether the sub-soil is aggressive or not, it will be prudent to use the M4 mixture for below DPC on all sites.

c). *Do not* use brickies loam.

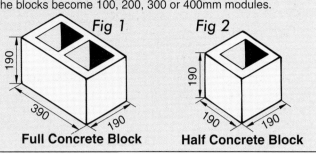

Masonry Block Modules
While the actual block dimensions are either 90, 190, 290 or 390, when the 10mm vertical and horizontal joints are added, the blocks become 100, 200, 300 or 400mm modules.

Fig 1 190, 390, 190 — **Full Concrete Block**

Fig 2 190, 190, 190 — **Half Concrete Block**

this stage are most vulnerable to bumps by careless handling of scaffolding, wheelbarrows etc.

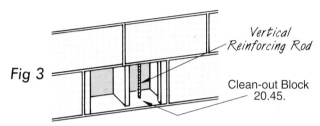

Fig 3 — Vertical Reinforcing Rod; Clean-out Block 20.45.

Clean out Mortar Droppings

Before pouring grout, all mortar droppings should be cleaned out of the vertical cores and bond beams. Any projecting mortar dags are knocked off using a reinforcing rod. Clean-out openings are provided at the base of the wall to remove the droppings (fig 3).

Grouting or Blockfill

Grout should have a compressive strength of 20MPa with cement content *not* less than 300kg per cubic metre. While small walls could be grout filled by hand pouring, for larger jobs it is best to use a pump. In very hot weather it may be necessary to hose the cores to cool the blocks to prevent 'flash setting' of the grout. This should be completed at least thirty minutes prior to pouring grout. For walls higher than 3m, grout fill in two stages about 30 minutes apart.

Footings & Slab for Block Construction

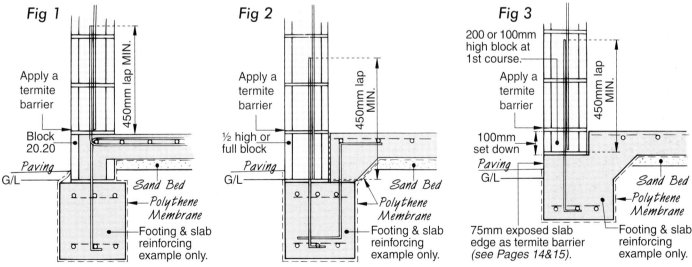

Fig 1 — Apply a termite barrier; 450mm lap MIN.; Block 20.20; Paving; G/L; Sand Bed; Polythene Membrane; Footing & slab reinforcing example only.

Fig 2 — Apply a termite barrier; 450mm lap MIN.; ½ high or full block; Paving; G/L; Sand Bed; Polythene Membrane; Footing & slab reinforcing example only.

Fig 3 — 200 or 100mm high block at 1st course.; Apply a termite barrier; 450mm lap MIN.; 100mm set down; Paving; G/L; 75mm exposed slab edge as termite barrier *(see Pages 14&15).*; Sand Bed; Polythene Membrane; Footing & slab reinforcing example only.

Fig 1 is a common slab edge detail in which 'Bond Beam 20.20' is utilised as the slab edge formwork. One side shell is broken out to allow the outer shell to become the formwork for the wet concrete *(see fig 4, Page 38).* 'Block 20.61' can also be used as edge formwork. Many designers still prefer to have a set-down rebate in the slab edge as in figs 2 & 3. The reasoning is that should water gain access to the raw block it is natural that some moisture would wick across the webs to the interior surface.

The set down provides somewhere for a small excess of water to go rather than across the floor as could occur in fig 1, Page 37. With both figs 1 & 2, the footing concrete is poured, the first block course laid then the slab floor laid utilising the first course as formwork. Fig 3 on Page 37 is a raft slab. The footings and floor are poured together.

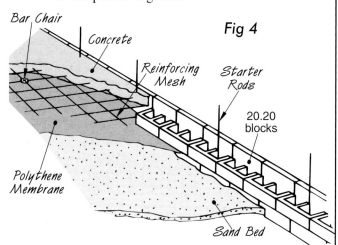

Fig 4

Embedding & Aligning Starter Rods

These are embedded at footing stage or when pouring the slab floor.

Step 1 A string line is stretched and attached to profiles to align with the centre of the block cores.

Step 2 The position of each rod is marked on the formwork or when installing into footings, where no formwork exists, peg a batten on the ground on which to mark each rod position.

Step 3 Hook cogged rods under footing reinforcement or embed 250mm MIN. Rods should extend above slab surface 450mm MIN. At this stage the rods will tend to fall over in the wet concrete. As the concrete firms their positions should be rechecked (see fig 5).

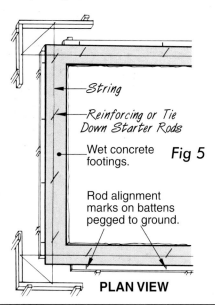

Fig 5

PLAN VIEW

Waterproofing: Walls must be painted to conform to the BCA. Clean walls down. Fill all holes. Apply three coats of approved 100% acrylic paint. *For alternative methods see block manuf. recommendations and the BCA.*

Block Masonry Wall Heights & Lintels

A wide variation in lintel heights and design exists from town to town. The block manufacturer will provide information on the locally adapted method.

The normal 2400mm ceiling height requires 12/200mm high blocks (see fig 1). In tropical areas it is common for ceiling heights to be 2600 and 2700mm to accommodate ceiling fans. The top course in most localities is a 'No. 20.20' knock-out bond beam course.

Above windows, a 'No. 20.12' or '20.25' lintel course is also required (see figs 2-5, Page 39). The standard door is 2100mm. Some situations will require a deeper lintel/bond beam as in figs 3-5. This may be a factor which will determine the ceiling height. In some localities, a three-quarter lintel 300mm high is used above these doors (see fig 3).

In some areas, the three-quarter lintel is used around the entire perimeter of the house. However, this will necessitate commencing at slab level with a half high block unless the ceiling height is increased to 2500mm. In some localities where 400mm high bond beams are required, the lintels are kept at 2000mm high and manufacturers provide special 2000mm high doors and frames.

Control Joints

Reinforced masonry walls which are less than 3m high and which also have a perimeter reinforced bond beam on the top course must *not* have control joints.

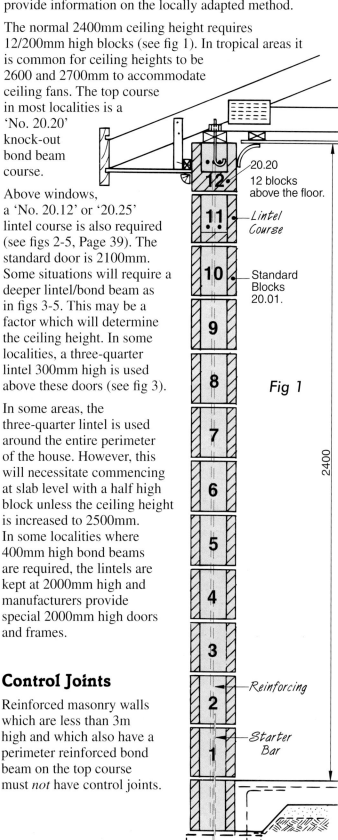

Fig 1

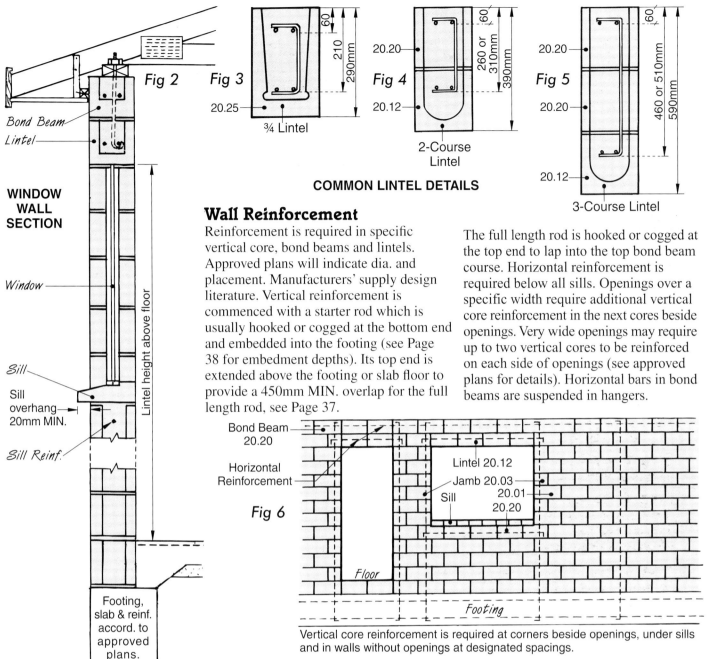

Fig 2

Fig 3
¾ Lintel
60
210
290mm

Fig 4
20.20
20.12
60
260 or
310mm
390mm
2-Course
Lintel

Fig 5
20.20
20.20
20.12
60
460 or 510mm
590mm
3-Course Lintel

COMMON LINTEL DETAILS

Bond Beam
Lintel

WINDOW WALL SECTION

Window

Sill

Sill overhang 20mm MIN.

Sill Reinf.

Lintel height above floor

Footing, slab & reinf. accord. to approved plans.

Wall Reinforcement

Reinforcement is required in specific vertical core, bond beams and lintels. Approved plans will indicate dia. and placement. Manufacturers' supply design literature. Vertical reinforcement is commenced with a starter rod which is usually hooked or cogged at the bottom end and embedded into the footing (see Page 38 for embedment depths). Its top end is extended above the footing or slab floor to provide a 450mm MIN. overlap for the full length rod, see Page 37.

The full length rod is hooked or cogged at the top end to lap into the top bond beam course. Horizontal reinforcement is required below all sills. Openings over a specific width require additional vertical core reinforcement in the next cores beside openings. Very wide openings may require up to two vertical cores to be reinforced on each side of openings (see approved plans for details). Horizontal bars in bond beams are suspended in hangers.

Bond Beam 20.20

Horizontal Reinforcement

Lintel 20.12
Jamb 20.03
20.01
Sill
20.20

Fig 6

Floor

Footing

Vertical core reinforcement is required at corners beside openings, under sills and in walls without openings at designated spacings.

Roof Tie Down

Various tie down methods are applied depending on the force to be resisted. Three common ones are illustrated. Additional methods can be found in manufacturers' literature. See approved plans for specific requirements.

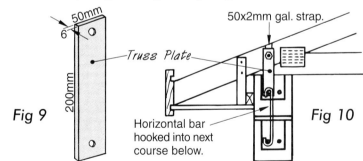

50mm
6
200mm

Fig 9

Truss Plate

50x2mm gal. strap.

Horizontal bar hooked into next course below.

Fig 10

Cyclone Areas

In cyclone areas, it is common practice to anchor roof trusses or rafters using mild steel plate 200x50x6mm with a hole at each end (see figs 9 &10). The bottom end is threaded over the bond beam reinforcing while the top end is bolted to trusses or rafters using 1/12mm bolt. The horizontal rod is anchored down into the next course below. Additional tie-down can be obtained by strapping the truss and bolting through the strap (see fig 10) and by applying the plate to both sides of the truss. Refer to AS4773.1 for details and capacities or truss plate.

Triple grips with 4/2.8mm gal. flat head nails in each leg and 2 to plate.

Top plate anchor bolts spaced accord. to approved plans.

30x.8mm gal. strap taken under top plate and fastened accord. to manuf. requirements.

Top Plate Anchor Bolts

Fig 7

Fig 8

TIE DOWN USING TOP PLATE

2

Concrete Footings & Slab Floors

Siting the House

Prior to setting up profiles or excavating a cut and fill site, the house site area must be cleared of trees and vegetation and the house sited on the allotment. This simply means the proposed corners of the house are pegged.

Steps to Siting a House

(see example fig 1)

Step 1 Stretch a string line through the front and side boundary pegs. A surveyor may be required to position new boundary pegs.

Step 2 Measure back the required distance from the front boundary given on the site plan to the proposed outside face of the house. Fix pegs calling them 'A' and 'B'. These pegs represent the width of the house and must at least be the legal minimum required distances from the side boundaries.

Step 3 Find peg 'C' by measuring from peg 'B'. To ensure it is square use 'Pythagoras Theorum'. Check to ensure it is at least the legal required minimum distance from the back boundary. Peg 'D' is found by measuring from 'A' and 'C'. Forget about the offset at this stage if the plan shows one. Roughly square these four pegs by taking measurements diagonally and adjust both rear pegs sideways until both diagonals are the same length. Alternatively, use the scientific calculator and the 'Triangulation Method' described more in the manual 'How to be a Successful Owner Builder & Renovator'.

On Irregular Sites

If the house is parallel to one boundary only: Square the building from this boundary ensuring again the legal minimum distances (as required by your Local Council) are maintained from all boundaries.

When the house is not parallel to any boundary:
Step 1 Establish one side, front, or rear of the house in the position desired. Ensure legal minimum distances between house wall or eaves and boundaries are maintained.

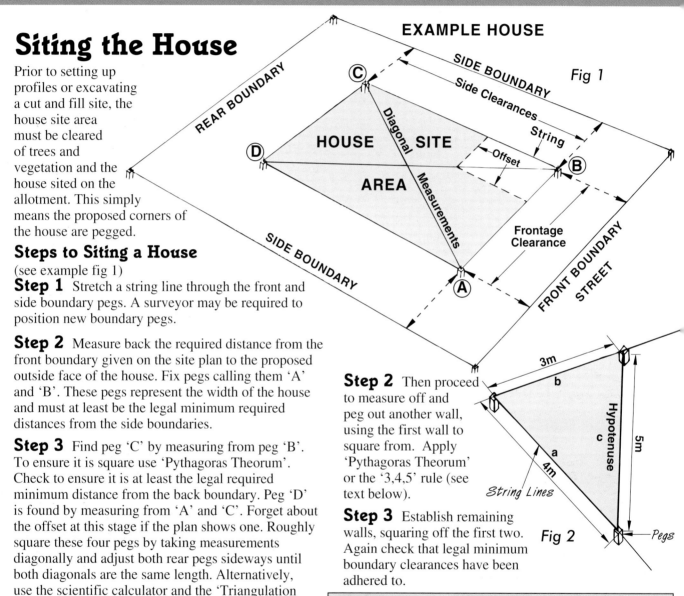

Step 2 Then proceed to measure off and peg out another wall, using the first wall to square from. Apply 'Pythagoras Theorum' or the '3,4,5' rule (see text below).

Step 3 Establish remaining walls, squaring off the first two. Again check that legal minimum boundary clearances have been adhered to.

HOW TO ESTABLISH SIDES AT RIGHT ANGLES

Pythagoras Theorem ($\sqrt{a^2+b^2}$) is the simplest, using fig 2 for example:

Square Side 'a' 4x4=16, then square Side 'b' 3x3=9, add them together 16+9=25, 25 is the square of side 'c'. Find the square of 25 by using the '$\sqrt{\ }$' button on the calculator.
5 becomes the length of 'c'.

The 3.4.5 Method of Squaring

A right angle can be established by the 3,4,5 method or their multiples such as 12,16,20 etc. Peg out the right angle roughly using the two smallest figures - 3 and 4m. Adjust them until the diagonal measures 5m. For accurate squaring of a large area, use greater multiples. It is important to hold the tape measure true and accurately to measure.

Cut & Fill Sites & Slab Floors

Footings and slab floors are designed according to the foundation sub-soil type. These are classified in the table opposite.

Cut & Fill Sites & Slab Floors

Many sites have some degree of fall or slope. Where concrete slab floors are to be laid, a bulldozer or backhoe is used to level the slope. Sometimes it is possible to level the entire house site area. However, on steeper slopes usually only a portion of the slab floor area can be levelled. The remainder is filled.

In the fill area the slab can be either: **'A'** Directly supported by compacted fill or **'B'** Be suspended on piers with uncompacted fill as in figs 1 & 2.

A. Slabs Directly Supported by Compacted Fill
Methods of Filling

Filling is required to be either *Controlled fill* or *Rolled fill*. Sand should *not* contain any gravel size material *(see AS 1289)*. Clay fill is required to be moist during compaction.

Controlled fill - Sand can be used to fill areas up to 800mm deep. It must be compacted in layers *not* more than 300mm thick by a vibrating plate or roller. Clay fill can be applied *not* more than 400mm deep and compacted in layers *not* greater than 150mm thick by a mechanical roller.

Rolled fill - Sand fill can be repeatedly rolled and compacted by an excavator but must *not* exceed 600mm deep and should be rolled in layers *not*

GENERAL CLASSIFICATION OF SITES	
CLASS	**FOUNDATION**
A	Most sand and rock sites with little or no ground movement from moisture changes.
S	Slightly reactive clay sites with only slight ground movement from moisture changes.
M	Moderately reactive clay or silt sites, which can experience moderate ground movement from moisture changes.
H	Highly reactive clay sites, which can experience high ground movement from moisture changes.
E	Extremely reactive clay sites, which can experience extreme ground movement from moisture changes.
A to P	Filled sites (for further details see AS 2870).
P	Sites which include soft soils; such as soft clay or silt or loose sands, landslip, mine subsidence, collapsing soils, soils subject to erosion; reactive sites subject to abnormal moisture conditions or sites which cannot be classified otherwise.

Note: For classes 'M', 'H', and 'E' further division based on the depth of the expected movement is required. For deep-seated movements, characteristic of dry climates and corresponding to a design depth of suction change 'H$_s$', equal to/or greater than 3m, the classification shall be M-D, H-D, or E-D as appropriate. For example, 'H-D' represents a highly reactive site with deep moisture changes, and 'H' represents a highly reactive site with shallow moisture changes.

The above table (Amendment No. 6) is taken from 'The Building Code of Aust.' by kind permission of the 'Australian Building Codes' board.

Note: A layer of clean quarry sand should be levelled over the top of the above fill to a depth of 20mm MIN.

greater than 300mm deep. Clay fill rolled with this method *should not* exceed 300mm deep and rolled in layers *not* exceeding 150mm thick. Further details should be obtained from the AS 2870 or a geotechnical engineer.

B. Slabs Suspended above Uncompacted Fill

The slab can be suspended over filled areas and supported by reinforced piers as in figs 1 & 2.

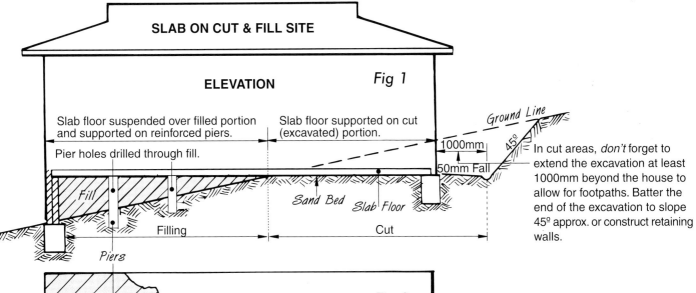

SLAB ON CUT & FILL SITE

ELEVATION *Fig 1*

Slab floor suspended over filled portion and supported on reinforced piers.

Slab floor supported on cut (excavated) portion.

Pier holes drilled through fill.

Ground Line

1000mm
50mm Fall

Fill

Filling

Sand Bed   Slab Floor

Cut

Piers

In cut areas, *don't* forget to extend the excavation at least 1000mm beyond the house to allow for footpaths. Batter the end of the excavation to slope 45° approx. or construct retaining walls.

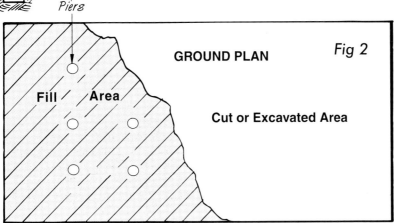

GROUND PLAN *Fig 2*

Fill Area

Cut or Excavated Area

In using option 'B', the suspended area of the slab is filled in order to support the reinforcing and wet concrete. This fill *does not* require compaction because after curing, the floor is fully supported by the piers and the perimeter wall. Both options should be designed by an engineer.

Setting out Profile Hurdles

For further details on 'Setting-out', see the manual 'How to be a Successful Owner Builder & Renovator'.

After the house area is sited and excavated (if necessary), profile hurdles are constructed to enable accurate levelling, squaring and aligning of the house on the site.

Erecting Profile Hurdles

Firstly drive in the four corner profile pegs keeping the tops above the finished floor level (if a slab-on-ground floor is to be used). Then with an optical or laser tripod level, place a level mark on each peg preferably at the slab surface level. *See Pages 15-17 for slab floor heights in relation to overflow relief gullies.* Fix a nail on these marks. Then string a line tight around the perimeter attached to these nails. This acts as a guide for fixing the remaining profiles. If the building has an offset such as an 'L' shape, use short sections of profiles at opposite sides of this offset both ways as in fig 1. Brace any profiles over 1200mm high and ensure profiles are rigid.

Squaring

Mark the front outside wall face alignment on the profiles. From this line, mark off the rear wall position and repeat the procedure for side walls. Then stretch a string line on nails to these points. Don't worry about offsets (if there are any) until these four positions are squared. Then to square the area within the string line, take diagonal measurements from the string line intersections as indicated by the dashed line in fig 1. Adjust the two rear nails supporting the side wall string lines. Adjust both the same amount until the diagonal measurements correspond. Keep in mind to maintain the legal distance measurement from boundaries. Offsets can be measured from these four fixed points.

Transferring Footing Lines to Ground Level

When the top surface of footings are below ground level, footings are usually wider than the walls they support and are centralised beneath those walls. However, when taken above ground level they are aligned (*as in fig 1&2, Pages 15&16*). If the site is *not* level, footing positions must be transferred to ground level. This is carried out by plumbing a line down to pegs sited directly below the footing marks on the profiles. Nails are driven on these marks. String lines are then attached and lime or plaster dust sprinkled beneath the lines to act as a guide for trenching equipment. Only the outside lines of the footings are required as trenching buckets are usually the correct trench width.

> **Materials Required:**
> Pegs 50x50mm or 70x35mm; levelling boards 100x25mm; braces 75x25mm or 75x50mm according to height; nails.

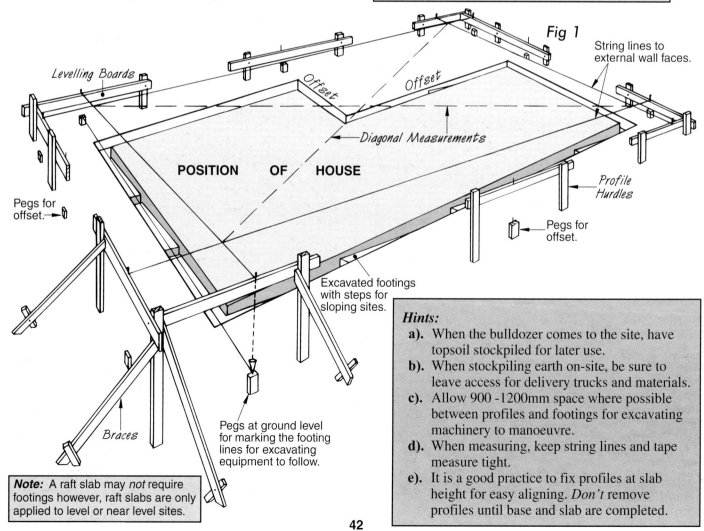

Fig 1

Levelling Boards

String lines to external wall faces.

Offset   Offset

Diagonal Measurements

POSITION OF HOUSE

Profile Hurdles

Pegs for offset.

Pegs for offset.

Excavated footings with steps for sloping sites.

Braces

Pegs at ground level for marking the footing lines for excavating equipment to follow.

> **Note:** A raft slab may *not* require footings however, raft slabs are only applied to level or near level sites.

> **Hints:**
> **a).** When the bulldozer comes to the site, have topsoil stockpiled for later use.
> **b).** When stockpiling earth on-site, be sure to leave access for delivery trucks and materials.
> **c).** Allow 900 -1200mm space where possible between profiles and footings for excavating machinery to manoeuvre.
> **d).** When measuring, keep string lines and tape measure tight.
> **e).** It is a good practice to fix profiles at slab height for easy aligning. *Don't* remove profiles until base and slab are completed.

Footing Preparation

After all footing lines are marked out, excavating can be carried out. This is best performed by a backhoe or chain digger. The builder should stay with the operator to instruct when the correct depths are reached then to check that the bottom of the trenches are level. Use a laser level or a straightedge and level for this. On sloping ground steps are cut at points which allow multiples of brick or block module courses to be laid (see figs 1 & 2). Corners of trenches are cut square and any soft spots in trench bottoms removed by spade. Remove all loose material.

Height Pegs or Laser Level

Footing heights can be established by an optical or laser level (a laser level can be operated without an assistant) during the concrete pour or by driving 25x25mm pegs along the sides of trenches 2400-3000 apart. Nails can be driven in the sides of the pegs to act as a guide for levelling the concrete. Where steps in footings occur, keep pegs in front of the excavated steps and drive nails to indicate the step heights.

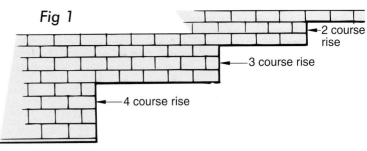

Fig 1

2 course rise
3 course rise
4 course rise

Brick Module Heights & Lengths
When bricks are to be laid in the footings, the footing risers and lengths should equal brick module lengths and heights including the 10mm joints.

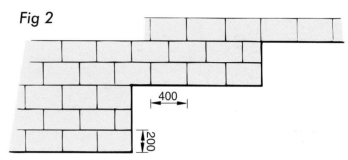

Fig 2

400
200

Concrete Masonry Block Module Heights & Lengths
Where masonry blocks are laid in footings, the footing lengths and risers should be in either 400, 300, 200 or 100 module lengths and heights. These include 10mm for joints.

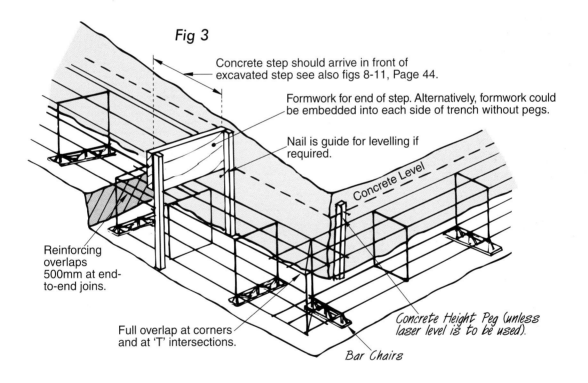

Fig 3

Concrete step should arrive in front of excavated step see also figs 8-11, Page 44.

Formwork for end of step. Alternatively, formwork could be embedded into each side of trench without pegs.

Nail is guide for levelling if required.

Concrete Level

Reinforcing overlaps 500mm at end-to-end joins.

Full overlap at corners and at 'T' intersections.

Concrete Height Peg (unless laser level is to be used).

Bar Chairs

Footing Reinforcement Alternatives

It is easier to obtain footing reinforcement cages in a prefabricated form than to fabricate on-site. This could be layers of trench mesh as in fig 4, Zed cages (fig 5), 'ⴖ' cages (fig 6) or whatever is specified and locally available.

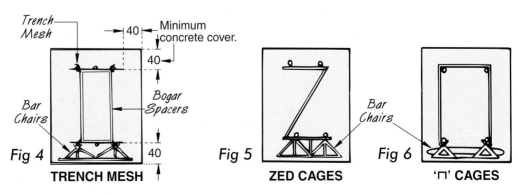

Trench Mesh
40
Minimum concrete cover.
40
Bogar Spacers
Bar Chairs
40
Fig 4
TRENCH MESH

Bar Chairs
Fig 5
ZED CAGES

Bar Chairs
Fig 6
'ⴖ' CAGES

Installing Reinforcing

The AS 2870 specifies a 40mm MIN. concrete cover to footing reinforcement at bottom, sides and ends. It is almost impossible to secure the reinforcement to an exact cover minimum during the concrete pour. For this reason, it is a good practise to increase the 40mm MIN. allowance. Ensure end-to-end joins on trench mesh have 500mm laps and corners are fully lapped. Support reinforcing on bar chairs at regular intervals not on bricks, stones or wood. Install form work in front of all steos (see fig 4, Page 43). Concrete can the be poured. Check while pouring that the 40mm MIN. clearance for reinforcing is maintained. Level off concrete using a straightedge. If vertical starter rods are required, position these in the wet concrete to a string line stretched between profiles (see fig 5, Page 38) unless they have already been tied into footing reinforcing.

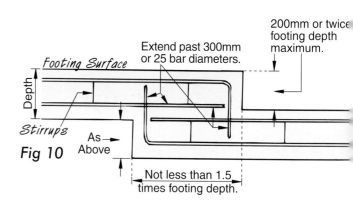

Fig 8
Footing Surface
Depth
200mm MAX.
Stirrups
As Above
Not less than 1.5 times footing depth.
Trench mesh cages bent up steps.

Fig 9
Footing Surface
Depth
200mm MAX.
Stirrups
200mm MAX.
Not less than 1.5 times footing depth.

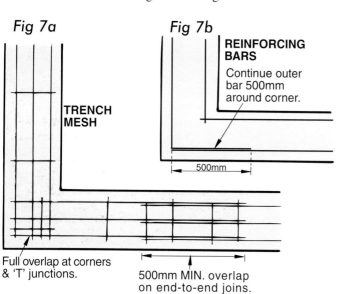

Fig 7a
TRENCH MESH
Full overlap at corners & 'T' junctions.

Fig 7b
REINFORCING BARS
Continue outer bar 500mm around corner.
500mm
500mm MIN. overlap on end-to-end joins.

Fig 10
Footing Surface
Depth
Stirrups
As Above
Extend past 300mm or 25 bar diameters.
200mm or twice footing depth maximum.
Not less than 1.5 times footing depth.

Notes: a). Water *should not* be permitted to lay in the bottoms of footings for long periods. Excavate any soft areas prior to pouring concrete.
b). Wet concrete in footings should be vibrated mechanically or by hand.

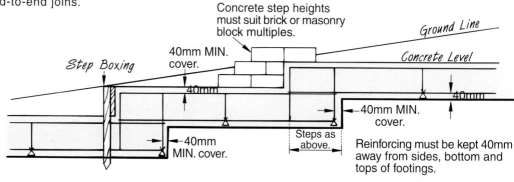

Fig 11
Concrete step heights must suit brick or masonry block multiples.
Ground Line
Concrete Level
40mm MIN. cover.
Step Boxing
40mm
40mm
40mm MIN. cover.
40mm MIN. cover.
Steps as above.
Reinforcing must be kept 40mm away from sides, bottom and tops of footings.

Drainage Pipes through Footings

Prior to pouring concrete, drainage pipes are installed through the middle third of footing trenches as in fig 12 and should be well lagged to allow for future footing movement. Where the pipe dia. is greater than the middle third of the footing, the footing depth should be increased.

Drainage trenches beside footings should be laid at the distances and depths from or below the footings as required in the BCA.

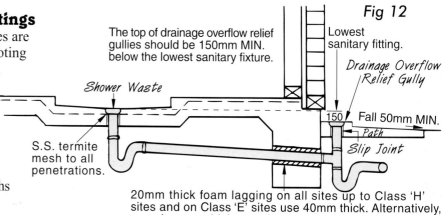

Fig 12
The top of drainage overflow relief gullies should be 150mm MIN. below the lowest sanitary fixture.
Lowest sanitary fitting.
Drainage Overflow Relief Gully
Shower Waste
150 Fall 50mm MIN.
Path
Slip Joint
S.S. termite mesh to all penetrations.
20mm thick foam lagging on all sites up to Class 'H' sites and on Class 'E' sites use 40mm thick. Alternatively, use sleeves which permit equivalent movement.

Problem Footings

Collapsing Ground —
A simple method of shoring is to use two lengths of sheet material (e.g. plywood). Four sheets are required if both sides are collapsing. The sheets are supported vertically at the sides of the trench while concrete is being poured.
A person back fills outside of the trench to the same heights as the rising concrete. As footings are filled, the first two sheets are continually repositioned (see fig 2). Fig 1 is sometimes necessary.
Use demolition timbers to save costs. Keep the bottom strut at ground level, thread reinforcing between struts and remove top strut as the concrete is being poured.

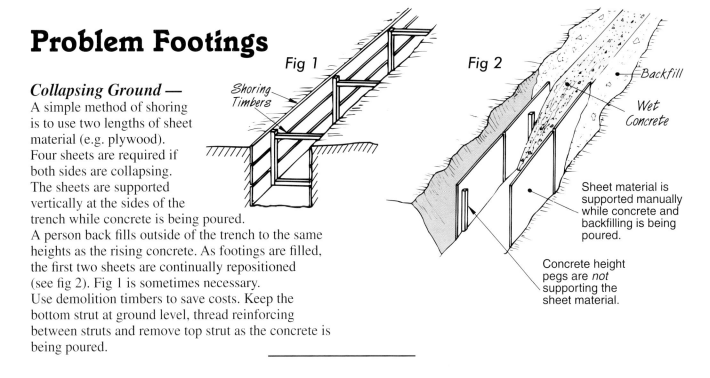

Fig 1 — Shoring Timbers

Fig 2 — Backfill — Wet Concrete — Sheet material is supported manually while concrete and backfilling is being poured. — Concrete height pegs are *not* supporting the sheet material.

Column or Stump Subfloors

Columns or stumps can be of steel tube, precast concrete or treated timber. Sectional size of steel, concrete or timber stumps will be specified in the approved plans as well as columns designated to be bracing sets and tie down. Tie down columns are required at specific spacings to anchor the building.

Steps in Construction
Step 1 Set out profiles as in fig 1, Page 42. Nails are driven to indicate column centres. String lines are stretched both ways to form a cross over the centre of the columns. A plumb bob is then dropped from this cross and a peg driven.

Step 2 Holes are dug or drilled to the dimensions specified on approved plans.

Step 3 Concrete is poured in the bottom of holes to the depth indicated on plans.

Step 4 Columns are lowered into holes and aligned to string lines stretched between profiles indicating their sides. Column footings are backfilled with concrete as in fig 2 and those required for tie down have reinforcing included as in fig 3.

Fig 2 — Joist — Bearers — Anchorage of floor structure to column accord. to plans. — Columns — Concrete Backfill — Concrete Pad — As Specified — **FOOTING TO COLUMNS NOT REQUIRED FOR TIE DOWN**

Fig 3 — As Specified — **FOOTING TO TIE-DOWN COLUMNS**

Fig 4 — Stud Wall — Bearers — Ant Capping — **Cantilevered Floor** — All column footing and reinf. to be accord. to approved plans.

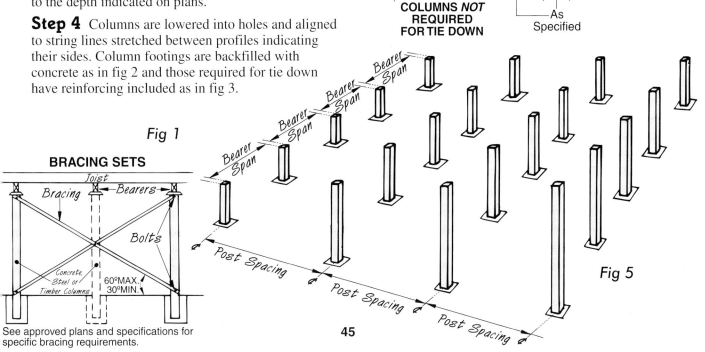

Fig 1

BRACING SETS

Joist — Bracing — Bearers — Bolts — Concrete, Steel or Timber Columns — 60°MAX. 30°MIN.

See approved plans and specifications for specific bracing requirements.

Bearer Span — Post Spacing

Fig 5

Pier & Beam Footings

Note: Driven timber piles may be used as an alternative to concrete piers for pier and beam footings. Enquire at your state timber association.

Pier and beam footings are used when adequate load bearing soils or strata can only be found at a greater depth. They should be designed by an engineer.

These footings are ideal where houses have to be built on filled or unstable sites or where variations in the subsoil type exist across the site. Footings should not be founded partly on rock and partly on other materials as differential movement may occur.

Steps of Construction

Step 1 The centre of pier holes are indicated along the centre of the proposed beam footings with small pegs. These positions will also be stipulated on the plans together with the diameter of the holes.

A mobile auger truck or tractor is engaged to bore the holes. Holes are taken down until the design bearing strata is reached. Loose material is removed and concrete is then poured to piers as soon as possible after drilling and prior to excavating trenches for beams. The concrete is terminated at the proposed bottom of beams.

Step 2 Footing trenches for the beams are excavated to the depth and width indicated on the plans.

Step 3 Height pegs are driven along the sides of the trenches at approx. 2400mm spacings to indicate the specified top surface heights of the beams. Reinforcing is then suspended in position and the concrete poured.

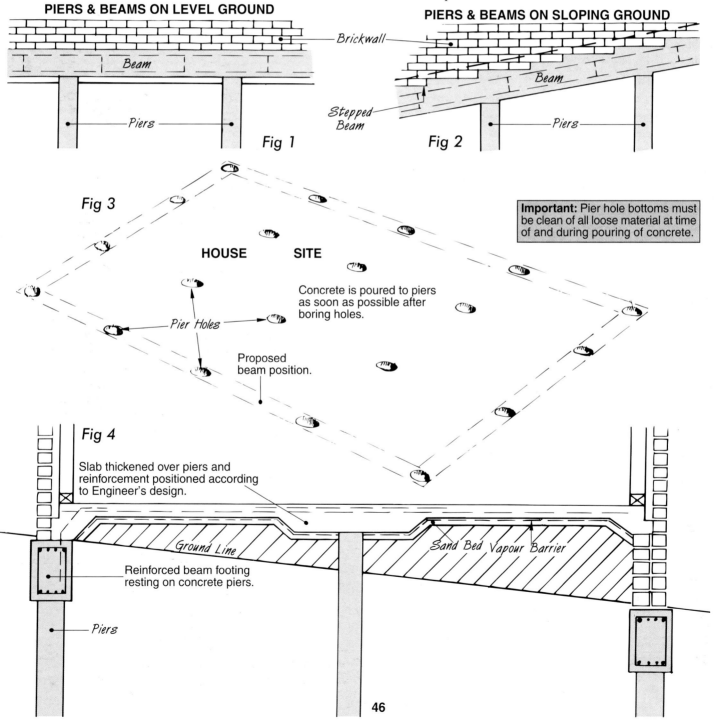

PIERS & BEAMS ON LEVEL GROUND

Brickwall
Beam
Piers

Fig 1

PIERS & BEAMS ON SLOPING GROUND

Brickwall
Stepped Beam
Beam
Piers

Fig 2

Fig 3

HOUSE   SITE

Concrete is poured to piers as soon as possible after boring holes.

Pier Holes

Proposed beam position.

Important: Pier hole bottoms must be clean of all loose material at time of and during pouring of concrete.

Fig 4

Slab thickened over piers and reinforcement positioned according to Engineer's design.

Ground Line

Sand Bed   Vapour Barrier

Reinforced beam footing resting on concrete piers.

Piers

46

Concrete Slab Floors

The surface of the slab should be a minimum height above finished ground level or paved surface. Plans should indicate or enquire at Local Authority; mimimum heights are required in flood prone areas. Also when establishing this height, keep in mind that the top of drainage overflow relief gullies (ORG) should be 150mm MIN. below the lowest sanitary fixture; most likely a floor waste *(see fig 1, Page 15 & fig 12, Page 44)*. They must also be clear of brick weep holes. Earth moving machinery is used to create a level site for the slab floor. This must be carried out prior to excavating footings. The ground at the perimeter should be graded to fall away from the slab at least 50mm for a distance of 1000mm around the perimeter of the slab.

Where compacted fill is required beneath the slab, *(see Page 41)*, check Local Authority requirements. You may opt to suspend the slab and use waffle pods as in Page 52, or reinforced isolated piers across the area as on Page 41. However, this may also necessiate an additional layer of slab reinforcing or a thicker slab or both.

Slab Floor On Ground Methods

There are variations in methods of slab floor construction.

Method A. A raft slab is formed and concreted with the footings to become a single entity. This method though requiring more effort to construct the formwork, enables the utilisation of exposed slab edges as a termite barrier as permitted in the BCA.

Method B. (Similar to 'C'). The slab is poured after and separate to the footings. *See alternative methods of forming the edge in figs 4-8.*

Method C. Slabs above sloping ground are supported underneath by brick or block walling and the mid spans by either compressed fill or isolated piers as on Page 41.

Formwork for Raft Slabs

Formwork will vary with the slab edge design.

Stages to Construct Raft Slabs

Stage 1 Level the site. If a footing is required, excavate at this juncture. Lay a 50mm thick layer of quarry aggregate across the site and compact.

Stage 2 Formwork is constructed and drainage pipes within the slab are laid as well as any termite barriers.

Stage 3 Lay the polyurethane membrane and waffle pods if required.

Stage 4 Reinforcing supported by bar chairs is installed.

Stage 5 Concrete is laid, compacted & cured.

Stages to Construct Slabs with Separate Footings

Stage 1 Excavate and prepare footings and pour the concrete.

Stage 2 Bricklayers are employed to bring any brickwork (or blockwork) up to floor level.

Stage 3 Install drainage pipes, electrical wiring or heating ducts etc. Erect perimeter formwork if required or use bricks as in figs 4 & 5.

Stage 4 Prepare basecourse and sand bed.

Stage 5 Apply any termite barriers required.

Stage 6 Place polythene membrane and lay reinforcing mesh with bar chairs.

Stage 7 Concrete is laid, compacted & cured.

Figure 1 is one method used for raft slabs; though most likely by a sub-contractor as the brackets need to be fabricated. Unless they can be hired.

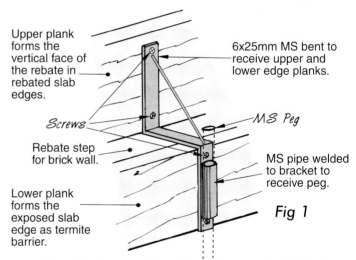

Upper plank forms the vertical face of the rebate in rebated slab edges.

6x25mm MS bent to receive upper and lower edge planks.

Screws

MS Peg

Rebate step for brick wall.

MS pipe welded to bracket to receive peg.

Lower plank forms the exposed slab edge as termite barrier.

Fig 1

When this formwork is utilised for a vertical slab edge termite barrier, the edge must be perfectly straight and smooth. Filling or rendering of the termite barrier face is *not* permitted *(see Pages 15&16)* for this reason standard dressed timber will only be useful on the first usage as bowing and cupping will likely prevent successful subsequent use. Laminated planks or steel sheet profiles are best for repeated use.

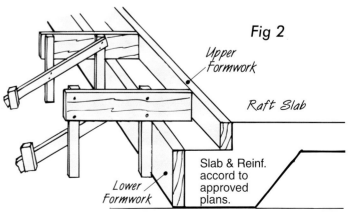

Fig 2

Upper Formwork

Raft Slab

Lower Formwork

Slab & Reinf. accord to approved plans.

Figure 2 also enables the slab edge to be used as a termite barrier. Figure 3, Page 48 is similar but care must be taken to vibrate under horizontal forms.

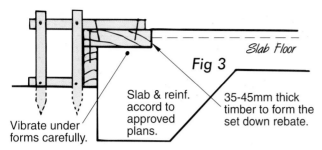

Fig 3

Slab Floor

Vibrate under forms carefully.

Slab & reinf. accord to approved plans.

35-45mm thick timber to form the set down rebate.

RAFT SLAB EDGE FORMWORK

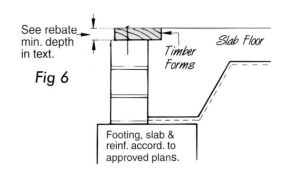

See rebate min. depth in text.

Fig 6

Slab Floor

Timber Forms

Footing, slab & reinf. accord. to approved plans.

Formwork for Separate Footings & Slab

It is good practice to make the depth of the rebate at least one full brick (see figs 4&5). This also enables bricks to be temporarily utilised as formwork as in fig 4 using weak mortar of 8:1 mix. The bricks can be later cleaned and reused. Alternatively, 145x45mm or 145x35mm timber could be fastened to the top of the brickwork extending into the proposed slab to form the rebate (see fig 6). The slab edge is sometimes formed up as in fig 7 or using other available steel or timber formwork. Where the footing is required to be extended above ground and the slab poured separately as in fig 9, formwork is applied to both sides of the footing after excavations are complete.

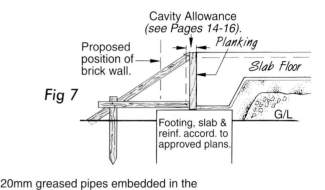

Cavity Allowance (see Pages 14-16).

Proposed position of brick wall.

Planking

Slab Floor

Fig 7

Footing, slab & reinf. accord. to approved plans.

G/L

20mm greased pipes embedded in the footing to be removed later. Holes remaining must be filled with grout to prevent entrapment of water and consequent corrosion of any footing reinforcement which may be exposed.

Fig 8

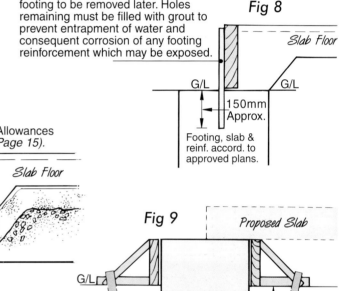

Slab Floor

G/L

G/L

150mm Approx.

Footing, slab & reinf. accord. to approved plans.

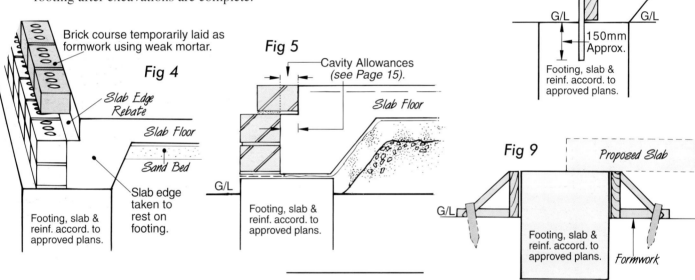

Brick course temporarily laid as formwork using weak mortar.

Fig 4

Slab Edge Rebate

Slab Floor

Sand Bed

Slab edge taken to rest on footing.

Footing, slab & reinf. accord. to approved plans.

Fig 5

Cavity Allowances (see Page 15).

Slab Floor

G/L

Footing, slab & reinf. accord. to approved plans.

Fig 9

Proposed Slab

G/L

Footing, slab & reinf. accord. to approved plans.

Formwork

Preparing to Pour the Concrete

Preparing Basecourse/Sand Bed

After constructing the formwork, stretch a string line across the top edge of the formwork from one side of the building to the other.
Make sure there is sufficient depth below this line to include both the basecourse/sand and concrete.
The thickness will be indicated on the plan. Areas that are too low, fill with gravel or metal. Top dress with sand to provide a soft bed to receive the polythene. The sand should be screeded off level with a straightedge. This will help to maintain the specified slab thickness all over the slab area and reduce concrete costs. The sand bed should be about 20mm deep.

Fig 1

Form level screeds across the sand with a straightedge. String lines stretched across the tops of the formwork will act as a depth guide. Measure down from these to ensure the concrete depth allowance.

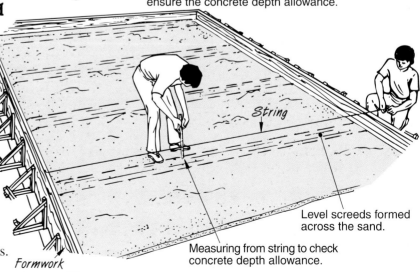

String

Level screeds formed across the sand.

Measuring from string to check concrete depth allowance.

Formwork

PREPARING FORMWORK & BASECOURSE / SANDBED

After forming the level sand screeds, they are used as a guide to level the infil.

Fig 2

Drainage Pipes & Services

At this stage the drainer should be employed to install any sanitary drainage. Drains may have to pass through the footings. If so, they should pass through the middle third of footings and be well wrapped with foam lagging *(see fig 12, Page 44)*. Where footings have been previously concreted then this work would have already been carried out and drains will only require extending to their slab surface positions. Any under slab electrical lines are also laid.

Laying the Polythene Vapour Barrier

Lay polythene vapour barrier to underlay the slab and edge beams and to extend out to ground level. Where the slab concrete is to be poured to adjoin with previously poured footings or where the slab and footings are to be poured together, check before laying the vapour barrier

that any sand or collapsed earth is cleared away from these areas. When the footings are to be poured with the slab as in raft slabs, the vapour barrier is laid prior to laying the footing reinforcement. Take special care that the reinforcement does *not* puncture the membrane. Allow 200mm laps at joins. Apply pressure sensitive tape to all joins and punctures and ensure a tight seal around pipes and fittings penetrating the polythene to prevent moisture rising.

Note: Vapour barriers are available in various weights, heavier ones providing longer life and being more resistant to puncture during construction. Check with approved plans to obtain the specified product.

Fig 3

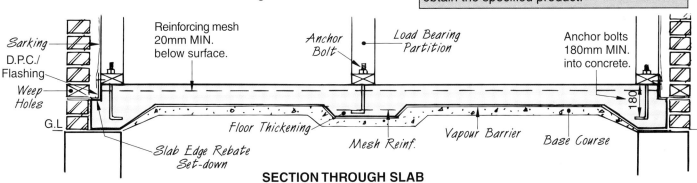

SECTION THROUGH SLAB

Slab Floor Reinforcing

Slab floor reinforcing is in the form of welded fabric mesh sheets 6000x2400mm or 1000x2400mm. The reinforcing is supported 20mm MIN. below the top surface of internal floors. This is to help reduce surface cracking. Where a second layer is required, it is supported 30mm above the plastic membrane.

Mesh in Exposed Slabs

Where slabs such as patios are exposed to weather, salt or pollution, then the top reinforcing is taken to 40mm below the surface. The corresponding min. strength for concrete is 32MPa. Check approved plans and specifications for specific requirements. Concrete cover measurements should be taken from the top of the uncoated portion of plastic coated bar chairs.

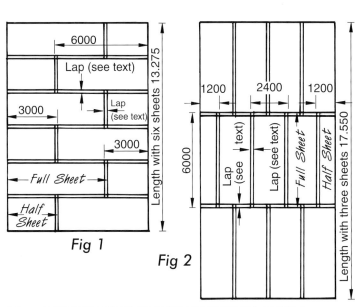

Placing the Mesh

It is important the mesh is maintained in the same level plane through the slab. This is carried out by the use of bar chairs of the correct height with bases and spaced at *not* more than 800mm ¢ or approx three chairs per sq.metre. Additional chairs will help prevent deformity to the mesh when tradespeople are walking over the surface.

The bases prevent the bar chair legs from puncturing the vapour barrier. Each sheet of mesh should be lapped over adjoining sheets in order for the tensile strength to be transferred from one to the other. The two outer transverse wires of one sheet must overlap the two outer transverse wires of the adjoining sheet at sides and ends. Laps should *never* comprise of more than two layers of mesh and no more than three at corners. To ensure this, a scale plan should be made of the proposed placement of sheets.

Place reinforcing mesh without puncturing the vapour barrier. Tie mesh together with wire or bag ties 500mm apart along joins. Bar chairs with their bases should then be placed. Any electrical wiring for under-floor heating is then laid.

Preparing to Pour the Concrete

Concrete *should not* be wheelbarrowed directly over the mesh. For this purpose, planks should be supported above the mesh to prevent the mesh from being deformed. This problem can be overcome by engaging a concrete pump to place the concrete. This is the ideal method as it also dispenses with the need for extra wheelbarrows and the cost of additional labour.

Mark Anchor Bolt/Starter Bar Locations

Mark bottom plate anchor bolts or starter bar positions on formwork prior to pouring slab or tie them to slab mesh.

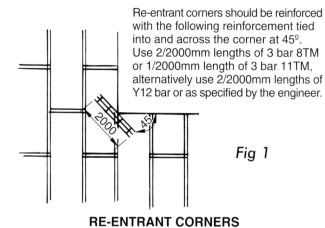

Re-entrant corners should be reinforced with the following reinforcement tied into and across the corner at 45º. Use 2/2000mm lengths of 3 bar 8TM or 1/2000mm length of 3 bar 11TM, alternatively use 2/2000mm lengths of Y12 bar or as specified by the engineer.

Fig 1

RE-ENTRANT CORNERS

Make a Check List

The day prior to pouring the concrete, make a check list of all items required for the pour including hardware, labour and tools, check the plans and specifications.

It is vitally important that items which are to be included or embedded in the wet concrete be on-site ready. Otherwise tradespersons are likely to ignore their inclusion. Remember that some of these items may need to be ordered days or weeks before.

Ordering Concrete

Three facts should be stated when ordering concrete:

1. *Quantity*

2. *Strength* normally 20-25 MPa (approved plans will specify).

3. *Slump* normally 100mm or pump mix

The strength should be obtained from the approved house plans or specifications. The slump is the plastic state of wet concrete. The concrete ordered should have a nominal 100mm slump unless an Engineer specifies otherwise. It is advisable to employ an experienced team to pour and finish the slab in one day.

Pouring the Concrete

Concrete is poured either by pump or wheelbarrow. One or two people carry out the levelling while others place and spread the wet concrete.

Concrete Compaction

Compact the concrete in beams, footings and slab with a vibrator to remove air voids.

Install Anchor Bolts & Hardware

Install hardware, anchor bolts. Bolts and similar hardware should be installed within approx. fifteen minutes of pouring concrete and straightened as concrete firms (*see 'Tie Down', Page 89*).

Levelling the Concrete

Concrete is screeded level by firstly forming level pads in the wet concrete using a laser level then from these, levelling bands or screeds are made using a straightedge.

When screeding wet concrete, ensure a smooth flat surface as free as possible of ripples or honey combing. If necessary, work the screed in order to raise sufficient cement paste to enable the bull float or trowelling machine to perform a better finish with less effort. Before the concrete firms, make sure all concrete dribbles are washed off the face of any brick base walls or finished surfaces. As the concrete firms it is floated and trowelled with a trowelling machine.

Note: Concrete strength is increased by 20% when a vibrator is used. Vibrate faces of exposed concrete behind the formwork to prevent air voids or honeycombing. Particularly where exposed concrete is to be utilised as a termite barrier.

Check Embedded Hardware

While surface floating and trowelling is being performed, anchor bolts, starter bars and any other embedded hardware should be checked for straight, level or plumb. Ensure there is *no* concrete build up around bolts and that the strips where framing bottom plates are to be laid are quite flat without any bumps or ridges.

Curing the Concrete

Concrete must be cured i.e. stop *premature* evaporation to achieve designed strength.

Two commonly used methods:

a). Lay plastic sheeting over the surface, seal all joins and weigh down all the perimeter edges to prevent wind exposure. Leave in place for 7 days.

b). Curing compound can be sprayed on the day of the pour when hard enough to walk on. Follow label instructions.

Wall frame contruction may be carried out during the curing period but *not* the day following the pour.

Doorways

Where rebates occur for garage or other doorways, ensure the nosing trowel is being used where required.

> **Important:** Curing of the concrete must commence *no later than 3 hours* after finishing and must be continuous for the duration of curing. Wetting down the concrete intermittently is *not* curing, this will cause expansion and contraction of the concrete and increase the likely incidence of several types of shrinkage cracking. *(From CCA Aust).*

Ceramic Tiles on Slabs

Ceramic tiles *should not* be laid on new slab floors for a minimum of three months unless slab reinforcement is increased to F92 through the tiled areas or other measures taken. Check with engineer.

> **Note:** The brick base illustrated shows brickwork taken down to footing. This method is best limited to sloping site usage. On level sites it is best to consider utilising the designs in figs 1 & 2, Pages 15&16 for the reasons outlined there.

Fig 1

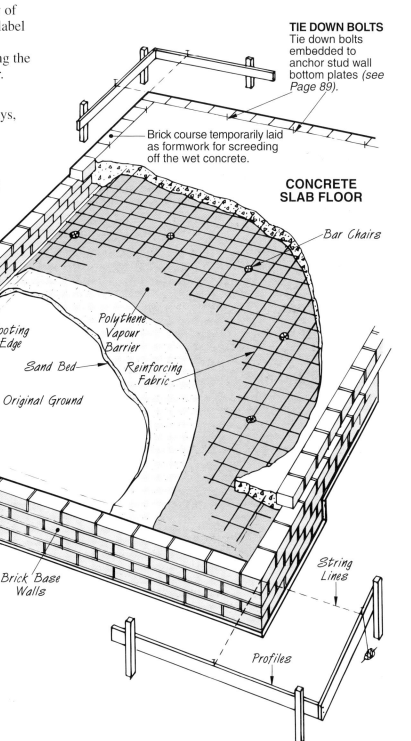

TIE DOWN BOLTS
Tie down bolts embedded to anchor stud wall bottom plates *(see Page 89).*

Brick course temporarily laid as formwork for screeding off the wet concrete.

CONCRETE SLAB FLOOR

Bar Chairs

Polythene Vapour Barrier

Footing Edge

Sand Bed

Reinforcing Fabric

Original Ground

Stepped Ground

Brick Base Walls

String Lines

Profiles

> **Hint:**
> It is a good practice to fix profiles at slab height. *Don't* remove profiles until base and slab are completed.

> **Note:** All pre-mixed concrete must be laid within one and a half hours of initial mixing. If using a retarder, check that it will be compatible with any adhesives required later for laying tiles or other materials.

The Waffle Pod & Slab System

The system is an on-ground one rather than in-ground so one major advantage is that excavated footings are *not* required. It is ususally used in conjunction with a raft slab and is ideal for reactive clay sites. All pods, internal ribs and edge beams, reinforcement and its placement are designed by an Engineer. The system utilises interlocking polystyrene formers. The formers are laid to create a gridwork of ribs or beams in which reinforcement is laid. Specially designed interlocking spacers ensure the polystyrene pods are locked together with the specified width ribs between. These spacers also support the reinforcement at the designated height.

Stages to Construct a Waffle Pod Slab

Stage 1 The site is levelled preferably by a laser guided drott. A 50mm deep layer of quarry aggregate is laid and compacted over the building area. Ensure this is completely level.

Stage 2 Drainage services within or below the slab are installed. A polythene membrane is then laid over the entire slab area.

Stage 3 The perimeter of the waffle pods are indicated by string lines and the pods are positioned.

Stage 4 Reinforcing is installed in the ribs and the slab mesh is then laid on bar chairs across the floor area 20mm below the finished slab surface or as specified. The concrete is poured and vibrated well especially in the ribs and edge beams.

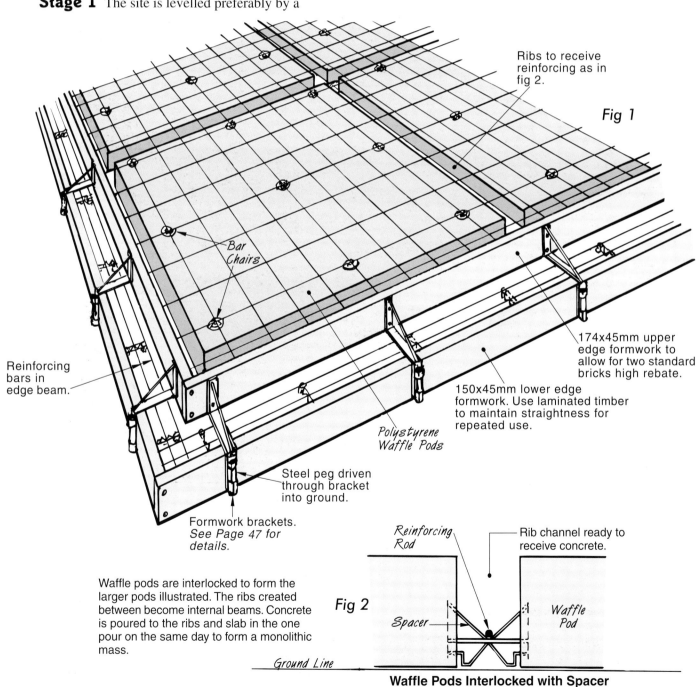

Ribs to receive reinforcing as in fig 2.

Fig 1

Bar Chairs

Reinforcing bars in edge beam.

174x45mm upper edge formwork to allow for two standard bricks high rebate.

150x45mm lower edge formwork. Use laminated timber to maintain straightness for repeated use.

Polystyrene Waffle Pods

Steel peg driven through bracket into ground.

Formwork brackets. *See Page 47 for details.*

Waffle pods are interlocked to form the larger pods illustrated. The ribs created between become internal beams. Concrete is poured to the ribs and slab in the one pour on the same day to form a monolithic mass.

Fig 2

Reinforcing Rod

Rib channel ready to receive concrete.

Spacer

Waffle Pod

Ground Line

Waffle Pods Interlocked with Spacer

3

Timber Floors & Subfloors

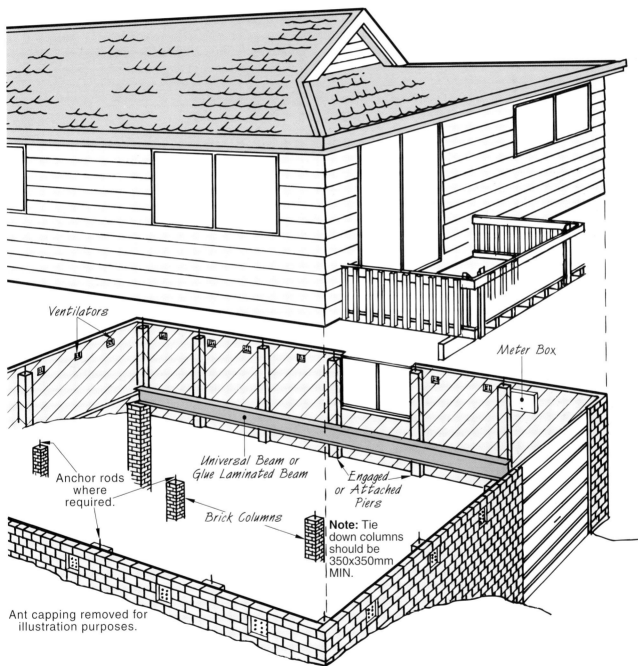

Ventilators

Meter Box

Anchor rods where required.

Universal Beam or Glue Laminated Beam

Engaged or Attached Piers

Brick Columns

Note: Tie down columns should be 350x350mm MIN.

Ant capping removed for illustration purposes.

Brick Base Construction
(see also Page 24).

Brick or masonry block base walls can be constructed on level or sloping ground. On steeper sites the lowest side can provide garage or storage space.

Stages in Construction
Stage 1 Footings with tie down starter rods are completed.

Stage 2 Brick walls and piers constructed with ventilators, windows (if required) and a meter box.

Stage 3 Anchor bolts and pier reinforcement grouted into brickwork where necessary.

Single Leaf Brick Bases

For designing single leaf masonry subfloor base walls see Page 23-25. Single leaf base walls usually have engaged piers either reinforced or unreinforced. These piers will in turn support the floor bearers. Reinforced engaged piers will of necessity have to be at least 350x350mm including the 110mm thick wall. Isolated piers are also 350x350mm; both in order to receive the reinforcement and grout fill.

Ventilators

Approved ventilators are positioned in walls immediately below bearer height. *See 'Ventilators' Page 24.*

Bearers & Joists

Bearers are the subfloor timbers laid on edge supporting the floor joists.

Bearer Span

The span is the distance between points of support. Bearers can be increased in sectional size in order to increase the span and in turn to reduce the number of piers or vice versa.

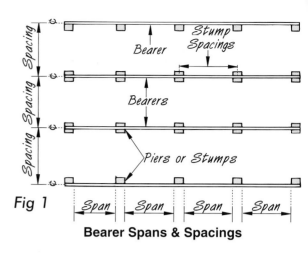

Fig 1

Bearer Spans & Spacings

Bearer Spacing

The spacing is measured between bearer centres and can be increased to reduce the rows of bearers. However, this in turn will require the joist depths to be increased. Reducing the spacing of the bearers may mean an extra bearer and a row of piers but the decrease in the joist depth will result in less shrinkage problems when using unseasoned joists.

Joist Shrinkage

A large unseasoned joist can shrink up to 15mm. This type of structural movement can cause a lot of future problems. It is advisable for all joists to be seasoned or kiln dried (Cypress has a low shrinkage factor). Alternatively, use an engineered flooring support system such as: hyJOISTs or similar.

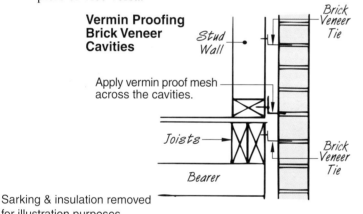

Vermin Proofing Brick Veneer Cavities

Apply vermin proof mesh across the cavities.

Stud Wall

Brick Veneer Tie

Joists

Brick Veneer Tie

Bearer

Sarking & insulation removed for illustration purposes.

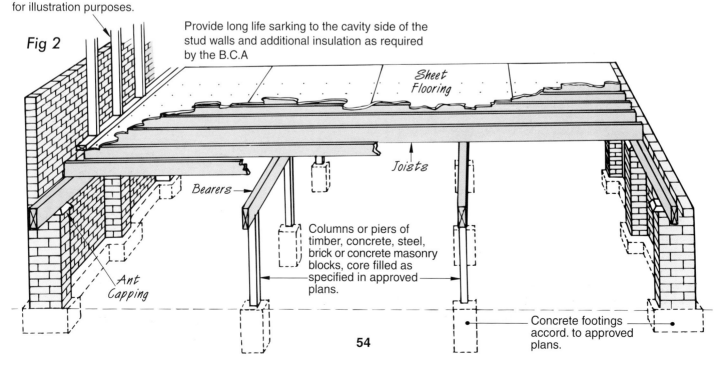

Fig 2

Provide long life sarking to the cavity side of the stud walls and additional insulation as required by the B.C.A

Sheet Flooring

Joists

Bearers

Columns or piers of timber, concrete, steel, brick or concrete masonry blocks, core filled as specified in approved plans.

Ant Capping

Concrete footings accord. to approved plans.

Laying Bearers

Ant Capping

Continuous ant capping must be built into the brickwork at the underside of framing members. See Page 65 for information on the type of metal to use and how the cappings should be made and laid.

Laying Bearers

Bearers should be laid in accordance with the AS 1684.2. They should be laid with any springs up and with a minimum of joins. Joins should only occur over supports. Ensure they are seated level and straight before anchoring.

If they are minimally below required height, they can be packed up with incompressible, decay resistant and non-corrosive material. Should they be minimally high, their bearing surface can be notched out.

Steps to Laying

Step 1 Check that the tops of base walls, columns or piers are terminating at the correct heights.

Step 2 Fit ant capping, D.P.C. or required flashings to tops of base walls or piers.

Step 3 Lay bearers on saw stools and square their ends. Prepare end joint scarfs or halvings if required (see figs 1 & 2).

Step 4 Mark off positions where any anchor bolts pass through the bearers then drill holes.

Step 5 Lay the outside bearers in position, check that their top edges are in level alignment with one another then nail together and bolt down. *For nominal fastenings see fig 6 to 13 on Page 56.*

When cutting joints in a bearer, ensure the two ends seat firmly and fit snugly together. Keep joints over the centre of support posts or stirrups. Allow 50mm MIN. bearing each side of a join.

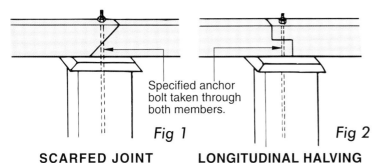

Specified anchor bolt taken through both members.

SCARFED JOINT *Fig 1*   **LONGITUDINAL HALVING** *Fig 2*

Step 6 Stretch string lines between the two bearers and install the intermediate bearers to the string lines. Badly sprung bearers may be partially cut over a support as in fig 3. However, a bearer cut in this manner must be reckoned from span tables as single span being supported at its end supports only. It pays to select reasonably straight bearers as they can be difficult to straighten.

Badly sprung bearers may be partially sawn through above support. A bearer sawn in ther manner must be considered as being supported at its end supports only (single span) regardless of its length.

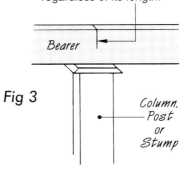

Fig 3

Bearer

Column, Post or Stump

STRAIGHTENING A SPRUNG BEARER

Anchoring Bearers

Anchoring methods will vary with the form of construction adopted. House plans will specify the fastening or anchoring method required. A house is only as secure as the anchorage provided so adhere to the specifications in case you ever have to make an insurance claim. Holes are drilled slightly over the bolt diameter used but *not* more than by 2mm in wood and 1mm in steel.

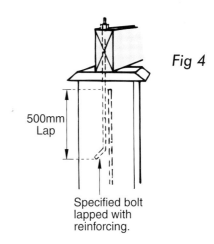

Fig 4

500mm Lap

Specified bolt lapped with reinforcing.

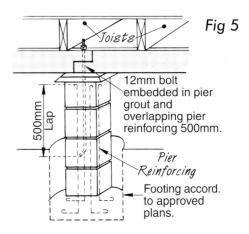

Fig 5

Joists

12mm bolt embedded in pier grout and overlapping pier reinforcing 500mm.

500mm Lap

Pier Reinforcing

Footing accord. to approved plans.

200x200mm Reinforced Block Masonry Piers

Nominal Joints & Fastenings for Bearers

Where specific fastenings are specified, it doesn't mean nominal fastenings can be omitted. Nominal fastenings are expected by the code to be applied as well in most instances unless overruled by the Engineer.

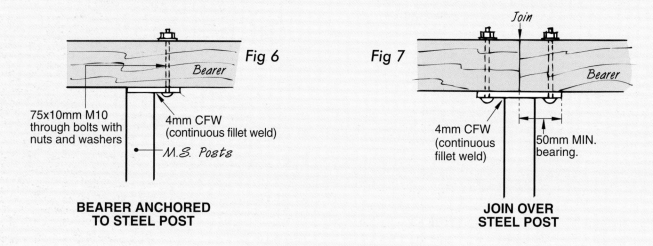

75x10mm M10 through bolts with nuts and washers

4mm CFW (continuous fillet weld)

M.S. Posts

Bearer

BEARER ANCHORED TO STEEL POST
(Fig 6)

Join

Bearer

4mm CFW (continuous fillet weld)

50mm MIN. bearing.

JOIN OVER STEEL POST
(Fig 7)

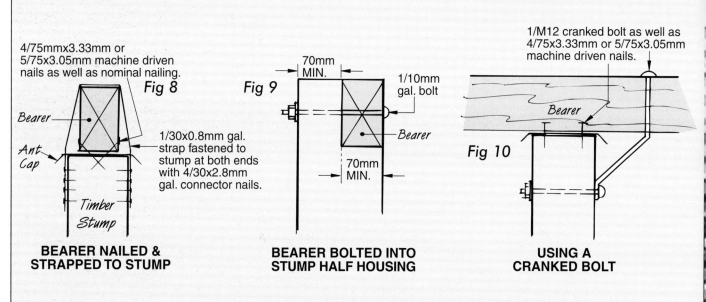

4/75mmx3.33mm or 5/75x3.05mm machine driven nails as well as nominal nailing.
Fig 8

Bearer

Ant Cap

Timber Stump

1/30x0.8mm gal. strap fastened to stump at both ends with 4/30x2.8mm gal. connector nails.

BEARER NAILED & STRAPPED TO STUMP

Fig 9

70mm MIN.

1/10mm gal. bolt

Bearer

70mm MIN.

BEARER BOLTED INTO STUMP HALF HOUSING

1/M12 cranked bolt as well as 4/75x3.33mm or 5/75x3.05mm machine driven nails.

Bearer

Fig 10

USING A CRANKED BOLT

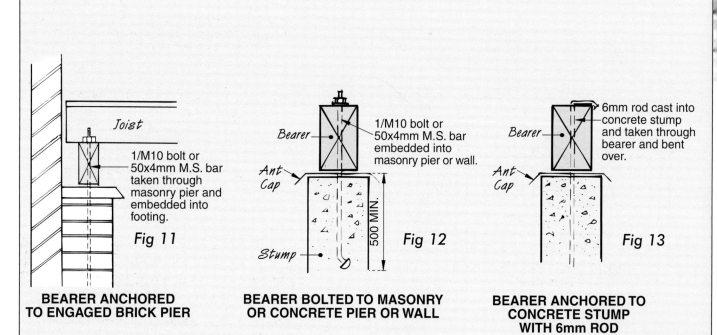

Joist

1/M10 bolt or 50x4mm M.S. bar taken through masonry pier and embedded into footing.

Fig 11

BEARER ANCHORED TO ENGAGED BRICK PIER

Bearer

Ant Cap

1/M10 bolt or 50x4mm M.S. bar embedded into masonry pier or wall.

500 MIN.

Stump

Fig 12

BEARER BOLTED TO MASONRY OR CONCRETE PIER OR WALL

Bearer

Ant Cap

6mm rod cast into concrete stump and taken through bearer and bent over.

Fig 13

BEARER ANCHORED TO CONCRETE STUMP WITH 6mm ROD

Floor & Roof Beams

Various types are available each having unique advantages and disadvantages see fig 1. It is vital that beams do not experience excessive shrinkage or spring causing a drop in height for the whole upper structure. This can result in binding joinery and out of alignment flooring, ceiling and roofing.

Ensure end bearing depth requirements are adhered to.

Installation

Prepare the same as you would for bearers working from a pair of sawstools. Before lifting into position, measure the beams depth and check if the top edge is going to arrive at the correct height. The piers may need raising or lowering. If raising, ensure packers consist of a rot proof and incompressible material. Ensure that all bolt holes have been pre-drilled and that the beam is straight. If it contains an even spring throughout its length, ensure the sprung edge is laid uppermost.

Fig 1

Beam Types & Their Advantages & Disadvantages

Solid Timber 
Can be purchased seasoned in sizes up to 240x90mm. Shrinkage can be a problem in unseasoned timber so request kiln dried. Before accepting delivery check for acceptable straightness.

Laminated Beams

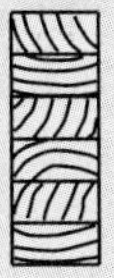

Laminated Veneered Lumber (LVL)

Manufactures can provide the design size. Laminated & LVL enables the use of smaller size beams or joists. These components are straight, stable and non-shrinking.
Remember you can purchase three grades of Laminated Beams:

1. *Structural* for where the beam is concealed.

2. *Standard* appearance Grade B for exposed but not sanded.

3. *Select* is same as standard but sanded and delivered wrapped in plastic.

Plywood Webbed Beam

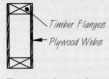
Timber Flanges
Plywood Webs

These can be purchased factory built or can be easily built on site. Their advantage is a weight reduction compared to other beams.

See ply manuf. construction spec. and tables for on-site construction

Mild Steel Beams 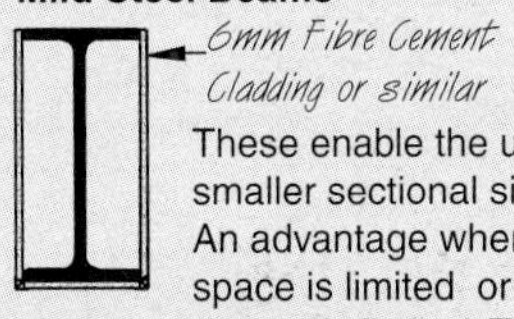
6mm Fibre Cement Cladding or similar
These enable the use of smaller sectional sizes. An advantage where space is limited or where head room is limited. These can be clad in fibre cement sheet or similar with PVC jointers at corners or adhere the sheet material using Liquid Nails and clamps or Contact Cement. Alternatively, apply framing then Gyprock line.

Floor Joists

Joist Types

Sawn timber is available in the sizes given in fig 2, however various benefits are derived from many of the alternative engineered joists. Almost all engineered joists are consistently straight, available in long lengths and suffer almost zero shrinkage. Some joists such as hyJOIST 'I' joists and Pryda Longreach floor trusses (figs 3 & 4) are easier to install services through. hyJOIST or similar are also extremely light weight.

Shrinkage Problems in Sawn Timber

Deep sawn hardwood and plantation softwood joists can present future structural problems if inadequately dry after the house is enclosed. Where these timbers are adopted, either: **a)**. Design the floor framing to enable the use of shallower joists, **b)**. Reduce bearer spans or **c)**. Utilise a higher stress grade. Kiln dried joists should always be preferred. Alternatively use Cypress which has a very low shrinkage factor.

SOLID SAWN TIMBER

Fig 2

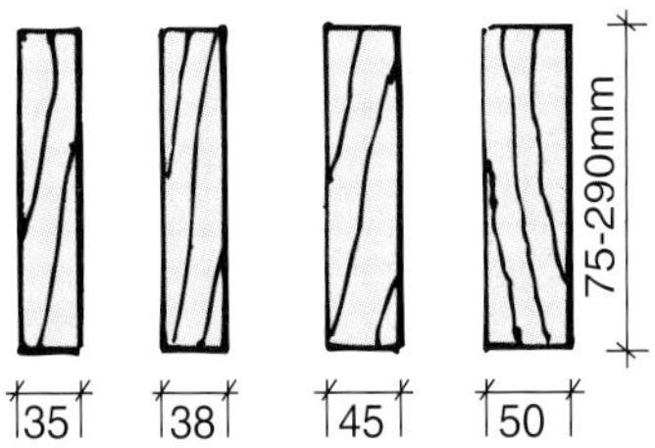

75-290mm

Depths vary between 75-290mm and lengths from 900mm increasing in 300mm increments.

35 38 45 50

Thicknesses vary between 35-50mm. 35&45mm thickness usually being gauged seasoned softwood and 38 and 50mm being sawn unseasoned hardwood or cypress.

Fig 3

hyJOIST ('I' Joists or similar)
Utilizes laminated Veneered Lumber (LVL)
These Joists arrive on-site true and straight and are lightweight. They suffer less shrinkage than solid timber joists and better utilize timber.

Fig 4

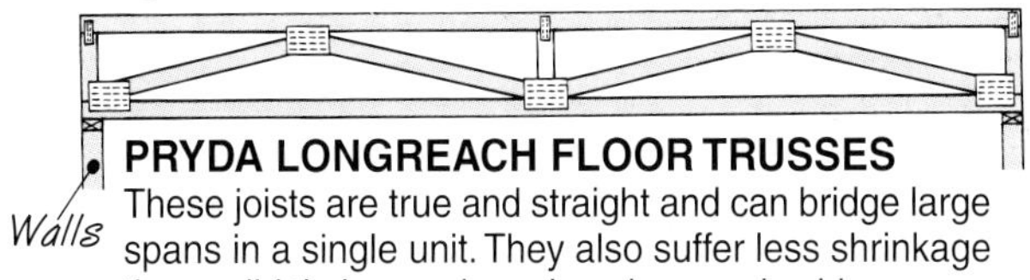

Walls

PRYDA LONGREACH FLOOR TRUSSES
These joists are true and straight and can bridge large spans in a single unit. They also suffer less shrinkage than solid timber and service pipes and cables are easily installed.

Notches, Holes & Cuts in Beams, Bearers, Joists & Rafters

For further information see AS 1684.2 & amendments. For notches, holes and cuts in engineered members refer to manufacturer's requirements.

The wrong size and/or placement of notches and holes in members can greatly reduce their designed strength. Notches are *not* permitted on both surfaces at ends. Where knots and defects are close by they should be regarded as holes and come under the requirements for hole spacings.

Notches & Cuts in Members

Notch in bottom of member at end over support

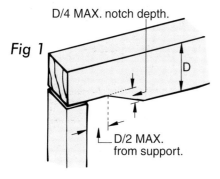

D/4 MAX. notch depth.

Fig 1

D

D/2 MAX. from support.

Notch in top of member at end over support

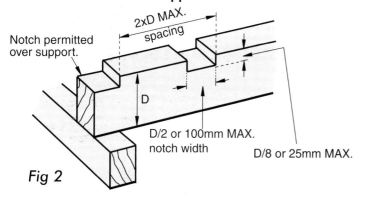

Notch permitted over support.

2xD MAX. spacing

D

D/2 or 100mm MAX. notch width

D/8 or 25mm MAX.

Fig 2

Spacing of Notches

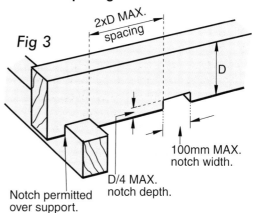

Fig 3

2xD MAX. spacing

D

100mm MAX. notch width.

D/4 MAX. notch depth.

Notch permitted over support.

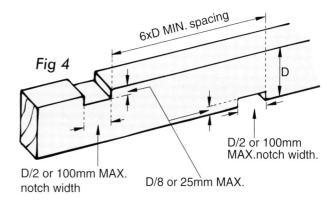

Fig 4

6xD MIN. spacing

D

D/2 or 100mm MAX. notch width.

D/2 or 100mm MAX. notch width

D/8 or 25mm MAX.

Holes in Members

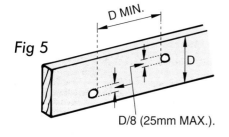

Fig 5

D MIN.

D

D/8 (25mm MAX.).

Three holes MAX. permitted per 1800mm of span

Spacing of Holes in Edges

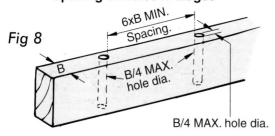

Fig 6

Hole within middle ⅓ of depth.

D/3
D/3
D/3

D

D/4 or 50mm MAX.

D less than 200mm

Only one hole permittted per 1800mm of span

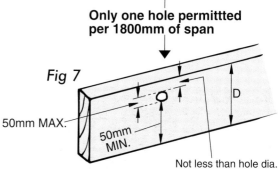

Fig 8

6xB MIN. Spacing.

B

B/4 MAX. hole dia.

B/4 MAX. hole dia.

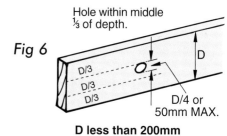

Fig 7

50mm MAX.

50mm MIN.

D

Not less than hole dia.

D greater than 200mm

Floor Joist Construction

There are two methods of floor joist construction, 'Platform' and 'Cut-in'(fitted). These are described in detail on Pages 60 & 61.

Floor joists are regularly spaced members supporting the flooring. Their cross sectional dimensions and spacings are given in the approved plans and specifications. Where joins occur, they must be over a support such as a bearer, beam or wall plate, see fig 4.

Fastening Joists

Joists should have 30mm MIN. bearing on bearers, beams or plates. Joists are fastened by either skew nailing, see fig 1, triple grips or joist straps using 30x3.15mm connector nails according to manufacturers requirements. See approved plans or specifications for specific requirement.

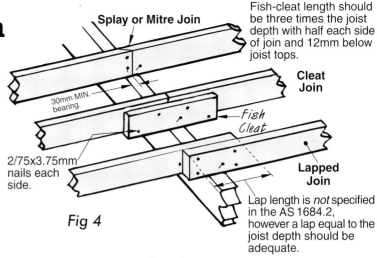

Fish-cleat length should be three times the joist depth with half each side of join and 12mm below joist tops.

Splay or Mitre Join

30mm MIN. bearing.

2/75x3.75mm nails each side.

Cleat Join

Fish Cleat

Lapped Join

Lap length is *not* specified in the AS 1684.2, however a lap equal to the joist depth should be adequate.

Fig 4

Methods of Joining Joists

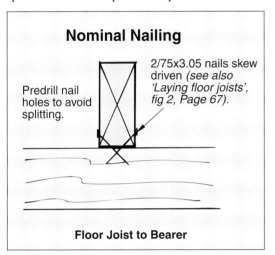

Nominal Nailing

Predrill nail holes to avoid splitting.

2/75x3.05 nails skew driven *(see also 'Laying floor joists', fig 2, Page 67).*

Floor Joist to Bearer

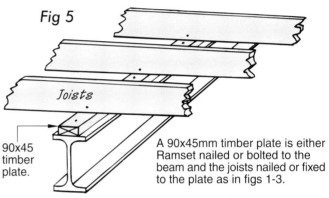

Fig 5

Joists

90x45 timber plate.

A 90x45mm timber plate is either Ramset nailed or bolted to the beam and the joists nailed or fixed to the plate as in figs 1-3.

Connecting Joists to Steel Beams

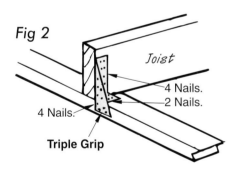

Fig 2

Joist

4 Nails.
2 Nails.

4 Nails.

Triple Grip

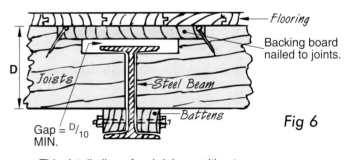

Flooring

Backing board nailed to joints.

D

Joists

Steel Beam

Battens

Gap = $^D/_{10}$ MIN.

Fig 6

This detail allows for shrinkage without causing a hump in the floor.

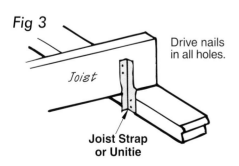

Fig 3

Joist

Drive nails in all holes.

Joist Strap or Unitie

Hint: It is a good time saver on the job if beforehand a plan is drawn to large scale of the proposed layout of the joists. This will also enable you to purchase the correct quantity of joists.

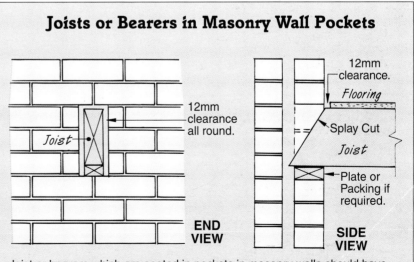

Joists or Bearers in Masonry Wall Pockets

Joist

12mm clearance all round.

12mm clearance.

Flooring

Splay Cut

Joist

Plate or Packing if required.

END VIEW

SIDE VIEW

Joist or bearers which are seated in pockets in masonry walls should have their ends splayed as above and 12mm MIN. space allowance all round.

Platform Floor Construction

Platform Flooring

This refers to flooring which is laid as a continuous membrane over the whole house area with walls being constructed on top of the floor. This method enables quicker and safer house construction, however it is *not* considered acceptable to use this method when laying tongue and groove timber strip flooring intended as a feature floor or polished floor.

When wet weather prevails during construction, holes are bored through the floor next to the bottom plates to prevent water ponding. These holes are concealed by

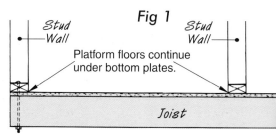

Fig 1

Stud Wall — Platform floors continue under bottom plates. — Stud Wall

Joist

the wall linings & skirtings.

Be sure to always purchase structural flooring grade particleboard which has a resin film on the upper surface. The side with the manufacturer's stamp indicating required joist spacing, should be laid face down.
For fixing plywood flooring see manufacturer's literature.

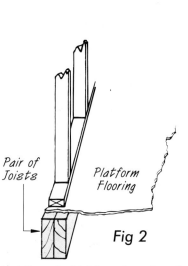

Fig 2

Pair of Joists — Platform Flooring

Install a pair of joists under external load-bearing walls and a single joist under non-load bearing external walls.

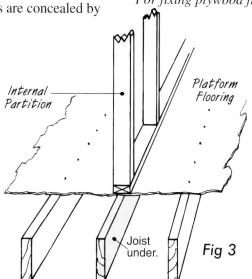

Internal Partition — Platform Flooring

Joist under.

Fig 3

Specifications may require one or more joists to support load bearing partitions or walls above. Check approved plans and specifications.

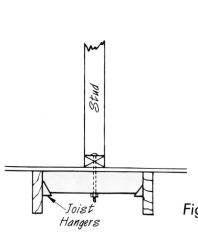

Stud

Joist Hangers

Fig 4

Where a roof load is concentrated on a stud, specifications may require solid blocking below the stud between the joists or an additional joist. Secure with joist hangers according to manufacturer's specifications.

JOIST LAYOUT FOR PLATFORM FLOORS

Intermediate blocking can be in line to support sheet joins as illustrated or staggered elsewhere for easier nailing.

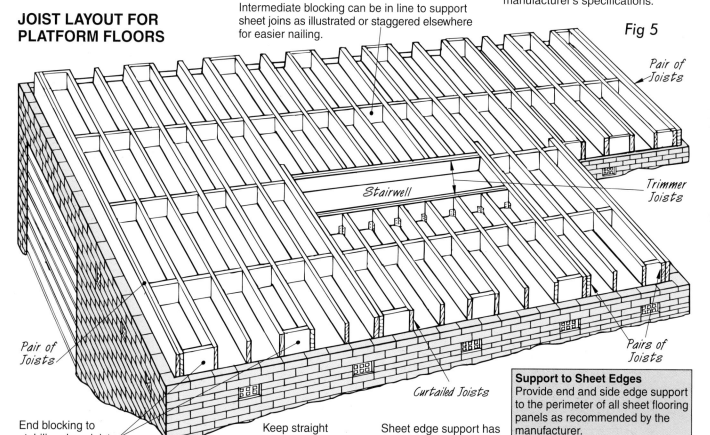

Fig 5

Pair of Joists

Trimmer Joists

Stairwell

Pair of Joists

Curtailed Joists

Pairs of Joists

End blocking to stabilise deep joists *(see Page 69)*.

Keep straight joists for sheet flooring joins.

Sheet edge support has been omitted for illustration purposes.

Support to Sheet Edges
Provide end and side edge support to the perimeter of all sheet flooring panels as recommended by the manufacturer.

Fitted Floors (or cut-in floors)

Cut-in flooring is laid after walls, partitions, roof and exterior cladding have been completed. They are cut-in to fit beside the wall bottom plates as in figs 1-5. Structural particleboard, plywood or T & G strip flooring or similar acceptable to the BCA may be used. A pair of joists is required under all walls and partitions parallel to joists. Flooring must have a 12mm MIN. bearing on the joists beside the wall bottom plates for nailing. End joins in joists should be made as in 'Methods of Joining Joists' fig 4, Page 59. Pairs of joists are spaced apart with a packer at 1200mm centres in order to provide the edge support for flooring. The packer is 25mm less in depth than the joists. This allows a 12mm space on the top and bottom edges of joists for shrinkage.

Solid Blocking

Solid blocking can be staggered for easier nailing as in fig 6 or in line for supporting a sheet join.

Two important factors should be considered:
a). Because the flooring is *not* laid until later, there is always the danger of falling between joists; and **b).** The joists are left exposed to the elements for a longer period than platform floors. Joists which are rain soaked may suffer swelling then shrinkage.

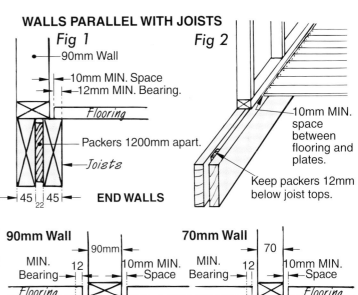

WALLS PARALLEL WITH JOISTS

Fig 1 — 90mm Wall, 10mm MIN. Space, 12mm MIN. Bearing. Flooring. Packers 1200mm apart. Joists. 45 22 45 **END WALLS**

Fig 2 — 10mm MIN. space between flooring and plates. Keep packers 12mm below joist tops.

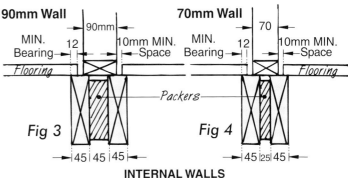

90mm Wall — MIN. Bearing 12, 90mm, 10mm MIN. Space, Flooring, Packers, Fig 3, 45 45 45

70mm Wall — MIN. Bearing 12, 70, 10mm MIN. Space, Flooring, Fig 4, 45 25 45

INTERNAL WALLS

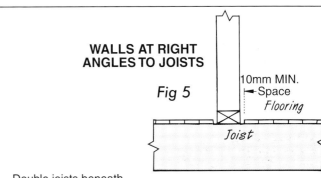

WALLS AT RIGHT ANGLES TO JOISTS

Fig 5 — 10mm MIN. Space, Flooring, Joist

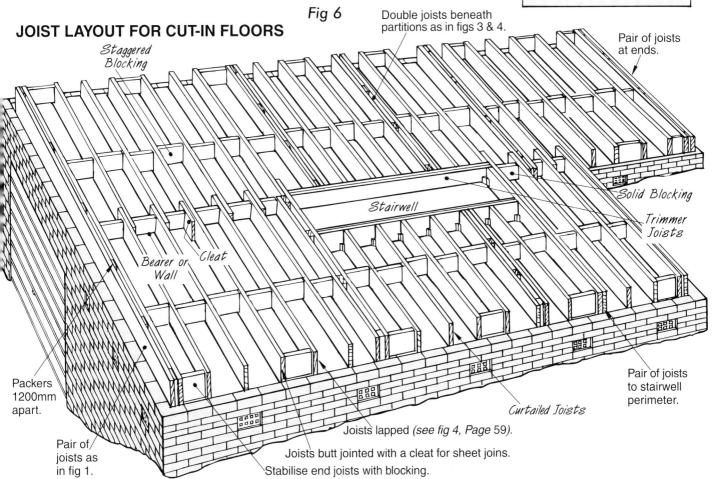

JOIST LAYOUT FOR CUT-IN FLOORS

Fig 6

Staggered Blocking

Double joists beneath partitions as in figs 3 & 4.

Pair of joists at ends.

Bearer or Wall

Cleat

Stairwell

Solid Blocking

Trimmer Joists

Packers 1200mm apart.

Pair of joists as in fig 1.

Pair of joists to stairwell perimeter.

Curtailed Joists

Joists lapped (*see fig 4, Page 59*).

Joists butt jointed with a cleat for sheet joins.

Stabilise end joists with blocking.

Using hyJOIST ('I' Joists)

Note: For tie down, hole cutting, cantilevers, openings in floors, racking through floors, comprehensive tables and additional information — *See manufacturer's literature.*

This floor joist system based on hyJOIST ('I' Joists) should be understood by all tradespeople concerned as its advantages certainly challenge traditional systems.

hyJOIST flanges are constructed from Hyspan 'LVL' (Laminated Veneered Lumber) and structural plywood is used for the webs. They offer benefits in efficient resource use, speed of installation, consistently straight edges plus negligible shrinkage.

The manufacturer's brochure contains the span tables for selecting the correct joists for the spans and loads. Lengths up to 13.2m are available. *Refer to the manufacturer's literature for further detailed designing and fixing instructions.*

The following details are solutions to some typical situations encountered on-site:

Blocking & Lateral Support

At Supports

Blocking or equivalent to prevent joists rolling at their supports is required and should be provided by one of the following methods:

1). Blocking with similar depth seasoned timber or hyJOIST offcut (see Detail F1).
2). Bracing with speedbrace or similar (see Detail F2).
3). Plywood closures combined with wall plates or battens (see Detail F3).

Single blocking can be arranged at 1800mm MAX. centres or in pairs at 3600mm centres in which case the joist adjacent to the outside joist should be blocked on both sides (or equivalent) and the spacing between pairs of blocking (or equivalent) along the supports is *not* to exceed 3600mm as shown in fig 2. In all cases the restraint systems must be linked to the tops of all adjacent joists by flooring or wall plates in the finished structure or by temporary battens during construction.

Between Supports

In the finished structure, with the flooring attached, no intermediate blocking or restraint is required unless to support sheet flooring edges. However, during construction prior to walking on bare joists, joists must be restrained at 2500mm MAX. ¢ with battens fixed back to points of rigidity.

Notching of hyJOISTS

Only limited notching according to the manufacturer's instructions is permitted.

Fixing of Flooring

Flooring may be nailed to joists using either 2.8mm diameter nails, hand driven or 2.5mm diameter nails, machine driven. Nail length should be minimum of 2.5 times the flooring thickness (*refer AS 1684.2*). Nail spacing should be at centres recommended for the particular flooring type. Where centres recommended are less than 50mm it is good practice to stagger their location so as to avoid the possibility of splitting. It is strongly recommended that flooring adhesive is used in conjunction with nailing.

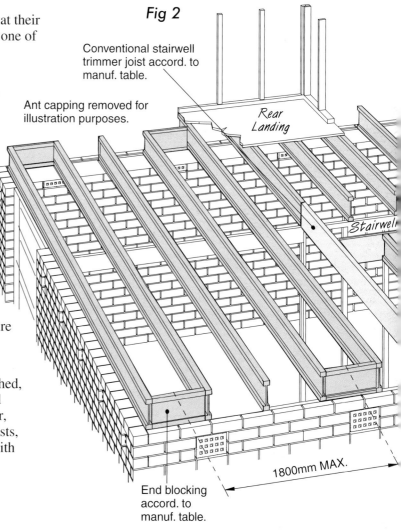

Fig 2

Conventional stairwell trimmer joist accord. to manuf. table.

Ant capping removed for illustration purposes.

Rear Landing

Stairwell

End blocking accord. to manuf. table.

1800mm MAX.

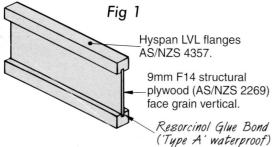

Fig 1

Hyspan LVL flanges AS/NZS 4357.

9mm F14 structural plywood (AS/NZS 2269) face grain vertical.

Resorcinol Glue Bond ('Type A' waterproof)

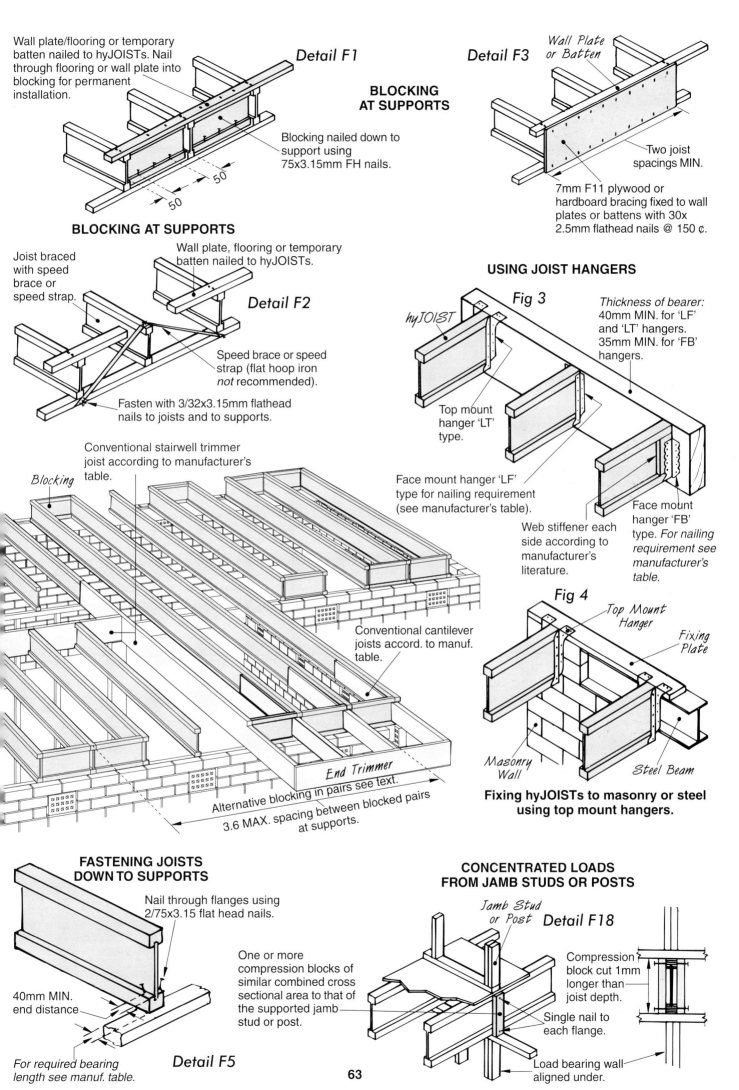

Detail F1

Wall plate/flooring or temporary batten nailed to hyJOISTs. Nail through flooring or wall plate into blocking for permanent installation.

BLOCKING AT SUPPORTS

Blocking nailed down to support using 75x3.15mm FH nails.

50
50

BLOCKING AT SUPPORTS

Detail F3

Wall Plate or Batten

Two joist spacings MIN.

7mm F11 plywood or hardboard bracing fixed to wall plates or battens with 30x 2.5mm flathead nails @ 150 ¢.

USING JOIST HANGERS

Detail F2

Joist braced with speed brace or speed strap.

Wall plate, flooring or temporary batten nailed to hyJOISTs.

Speed brace or speed strap (flat hoop iron *not* recommended).

Fasten with 3/32x3.15mm flathead nails to joists and to supports.

Fig 3

hyJOIST

Top mount hanger 'LT' type.

Face mount hanger 'LF' type for nailing requirement (see manufacturer's table).

Web stiffener each side according to manufacturer's literature.

Thickness of bearer: 40mm MIN. for 'LF' and 'LT' hangers. 35mm MIN. for 'FB' hangers.

Face mount hanger 'FB' type. *For nailing requirement see manufacturer's table.*

Blocking

Conventional stairwell trimmer joist according to manufacturer's table.

Conventional cantilever joists accord. to manuf. table.

End Trimmer

Alternative blocking in pairs see text.

3.6 MAX. spacing between blocked pairs at supports.

Fig 4

Top Mount Hanger

Fixing Plate

Masonry Wall

Steel Beam

Fixing hyJOISTs to masonry or steel using top mount hangers.

FASTENING JOISTS DOWN TO SUPPORTS

Nail through flanges using 2/75x3.15 flat head nails.

40mm MIN. end distance

One or more compression blocks of similar combined cross sectional area to that of the supported jamb stud or post.

For required bearing length see manuf. table.

Detail F5

CONCENTRATED LOADS FROM JAMB STUDS OR POSTS

Jamb Stud or Post **Detail F18**

Compression block cut 1mm longer than joist depth.

Single nail to each flange.

Load bearing wall aligned under.

63

Stairwell Trimmer Joists

Stairwell Trimmer & Trimming Joists

Refer to plans and specifications for cross sectional sizes of stairwell trimmers and trimming joists, as well as the fastening requirements.

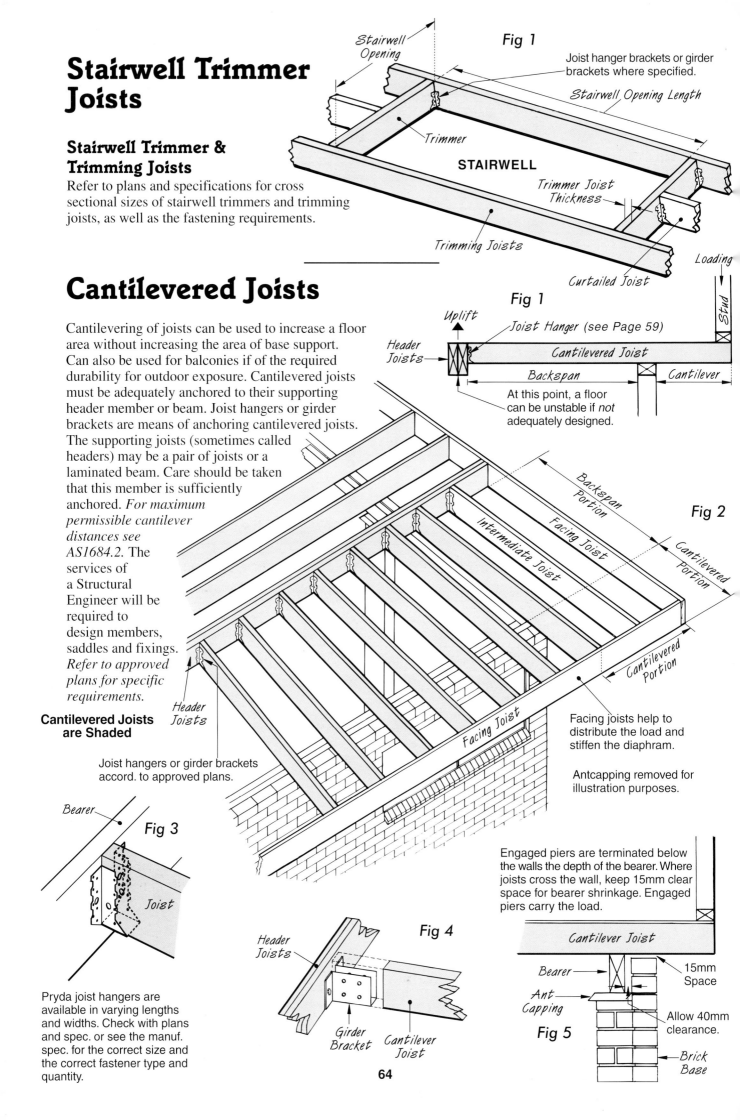

Fig 1

Stairwell Opening

Joist hanger brackets or girder brackets where specified.

Stairwell Opening Length

Trimmer

STAIRWELL

Trimmer Joist Thickness

Trimming Joists

Curtailed Joist

Loading

Cantilevered Joists

Cantilevering of joists can be used to increase a floor area without increasing the area of base support. Can also be used for balconies if of the required durability for outdoor exposure. Cantilevered joists must be adequately anchored to their supporting header member or beam. Joist hangers or girder brackets are means of anchoring cantilevered joists. The supporting joists (sometimes called headers) may be a pair of joists or a laminated beam. Care should be taken that this member is sufficiently anchored. *For maximum permissible cantilever distances see AS1684.2.* The services of a Structural Engineer will be required to design members, saddles and fixings. *Refer to approved plans for specific requirements.*

Cantilevered Joists are Shaded

Fig 1

Uplift

Joist Hanger (see Page 59)

Header Joists

Cantilevered Joist

Backspan

Cantilever

Stud

At this point, a floor can be unstable if *not* adequately designed.

Fig 2

Backspan Portion

Facing Joist

Intermediate Joist

Cantilevered Portion

Cantilevered Portion

Facing Joist

Header Joists

Joist hangers or girder brackets accord. to approved plans.

Facing joists help to distribute the load and stiffen the diaphram.

Antcapping removed for illustration purposes.

Fig 3

Bearer

Joist

Pryda joist hangers are available in varying lengths and widths. Check with plans and spec. or see the manuf. spec. for the correct size and the correct fastener type and quantity.

Fig 4

Header Joists

Girder Bracket

Cantilever Joist

Engaged piers are terminated below the walls the depth of the bearer. Where joists cross the wall, keep 15mm clear space for bearer shrinkage. Engaged piers carry the load.

Cantilever Joist

Bearer

Ant Capping

Fig 5

15mm Space

Allow 40mm clearance.

Brick Base

Ant Capping or Shielding (see also AS 3660.1)

Use 0.50mm MIN. thick gal. steel, or zincalume 0.40mm MIN. thick sheet copper, quarter hard or 0.50mm MIN. thick aluminium alloy. Stainless steel half hard 0.40mm MIN. Zincalume steel *can't* be welded without damaging the coating also bitumen coating will increase the life of the shields.

Joining

End Joins — Should be fully lock seam jointed, welded or riveted and soldered. ***External Mitre Joins*** — Should be fully soldered or brazed and welded. ***Internal Mitre Joins*** — Should have a gusset inserted and be fully soldered or welded or brazed.

Locating — Ant capping should be built into all brickwork and over all piers directly below the underside of bearers or framing which touches the subfloor supports. Ant caps are not required over sealed steel columns which are exposed on all sides. Capping over engaged piers must be joined to the wall shielding and must cover the full width of all masonry walls as in figs 4, 5 & 6. Where adjoining concrete floors are inaccessible, install shields as in fig 7. Turn edges down 45°. Edges should project 40mm MIN. from vertical faces of walls and piers (see fig 2).

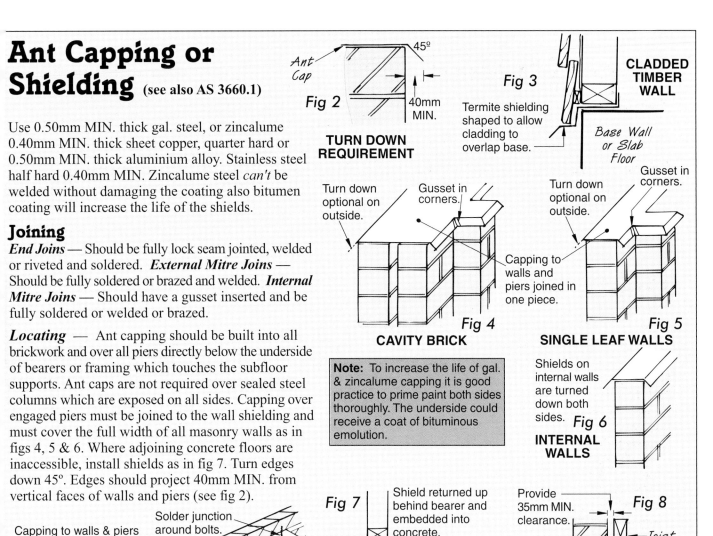

Fig 2 — TURN DOWN REQUIREMENT — Ant Cap — 45° — 40mm MIN.

Fig 3 — Termite shielding shaped to allow cladding to overlap base. — CLADDED TIMBER WALL — Base Wall or Slab Floor

Fig 4 — CAVITY BRICK — Turn down optional on outside. — Gusset in corners.

Fig 5 — SINGLE LEAF WALLS — Turn down optional on outside. — Gusset in corners. — Capping to walls and piers joined in one piece.

Fig 6 — INTERNAL WALLS — Shields on internal walls are turned down both sides.

Note: To increase the life of gal. & zincalume capping it is good practice to prime paint both sides thoroughly. The underside could receive a coat of bituminous emolution.

Fig 1 — Capping to walls & piers joined in one piece and taken across width of wall as in fig 4. — Solder junction around bolts.

Fig 7 — Shield returned up behind bearer and embedded into concrete. — Joist — Concrete Floor — Inaccessible Floor Area

Fig 8 — Provide 35mm MIN. clearance. — Joist — Bearer

Marking Joist Positions

All joist positions should be marked on the top of bearers, wall plates and beams prior to loading the joists. The following is a simple, quick and accurate method of setting out joist positions. A gauge block should be cut out of stock the thickness of which corresponds to that of the joists being used. *For example, a 90x45mm or 70x45mm for 45mm thick joists and 90x35mm for 35mm thick joists.* Then cut to length as in fig 1. To establish the joist spacings, hold gauge block square and parallel with the bearer and mark the joist spacing as in fig 2. To achieve accurate spacings, keep a sharp pencil point tight against the gauge.

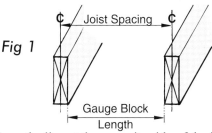

Fig 1 — Joist Spacing — Gauge Block Length

To show the line at the opposite side of the joist, simply use the gauge on its edge as in fig 3. An 'x' should be placed inside the lines to help clarify the joist position. Alternatively, the gauge block can be the above size plus one joist thickness and one side only can be marked to speed up the process. This is known as the **under and over** method.

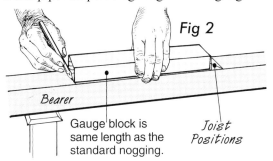

Fig 2 — Bearer — Gauge block is same length as the standard nogging. — Joist Positions

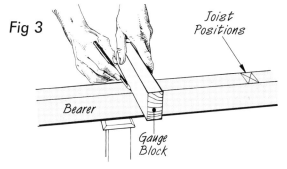

Fig 3 — Bearer — Gauge Block — Joist Positions

65

Setting Out Floor Joists

How to Mark Joist Positions

Step 1 From the house plan, find and indicate the positions of the external walls and internal load bearing walls that run parallel with the joists, marking them on the outside bearers and plates using light pencil lines (see fig 1).

Place the joist position marks above proposed wall positions and commence with any double joists. If there is a trimmed stairwell running parallel to the joists, indicate these trimmer joists. Then mark the position of joists supporting the edge of any sheet flooring joins.

Step 2 Work from one end and using the gauge block as described on Page 65, mark the intermediate joists. You will find when you arrive at a partition joist or joist supporting a sheet flooring join that

the spacing is shorter than the normal gauge length, simply centralise the last joist spacings or leave the last spacing shorter.

Step 3 Repeat these joist position marks onto the opposite side of the building. Then, holding a chalk line on these marks across the width of the floor, spring corresponding marks onto the tops of the intermediate bearers. Ensure joists *do not* cross the path of waste pipes such as where goosenecks arrive below bath and shower or WC outlets.

Procedure for Laying Joists

Step 1 Mark out positions of joists on bearers as below.

Step 2 Secure the joists in position. Double and trimmer joists first, then the remaining joists.

Step 3 Level off the top edges of the joists.

Step 4 Attach any solid blocking and noggings.

BEARER & JOIST LAYOUT

Note: For location of joists in relation to walls and partitions *(see Pages 60 & 61)*

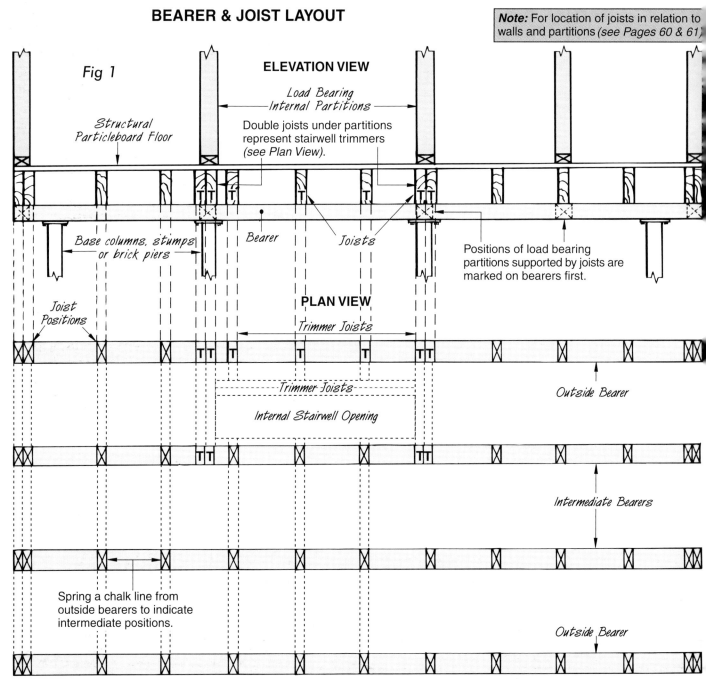

Fig 1

ELEVATION VIEW

Load Bearing Internal Partitions

Structural Particleboard Floor

Double joists under partitions represent stairwell trimmers *(see Plan View).*

Base columns, stumps or brick piers

Bearer

Joists

Positions of load bearing partitions supported by joists are marked on bearers first.

Joist Positions

PLAN VIEW

Trimmer Joists

Trimmer Joists

Internal Stairwell Opening

Outside Bearer

Intermediate Bearers

Spring a chalk line from outside bearers to indicate intermediate positions.

Outside Bearer

Laying Floor Joists

Laying Joists

Step 1 Joist ends are squared and cut to length. This procedure can be undertaken on sawstools at ground level. When the joists are later laid in their final positions, their top edges should be on the same level plane. If the joists have been brought to the site sized *(that is all dressed to the same depth)*, then continue with 'Step 2'. However, when rough sawn *(or unsized)* their bearing points will have to be gauged down to a uniform depth (see fig 1). This is known as 'cogging'.

How to Cog Joists to Equal Depths

Measure the depth of the narrowest joists and set your combination square to this depth. Using the square as a gauge, mark the ends of each joist with a carpenter's pencil. Mark to the bearer or plate width (see fig 1). Check out the excess. This waste could be from 1-6mm deep. Remember when marking, to gauge down from the sprung or upper edges of the joists.

Step 2 Join together all the double joists with packing pieces in place *(if they are required)* keeping the sprung edges uppermost. Working from one end proceed to skew nail the ends to the bearers.

If nails are splitting the joists, drill holes four-fifths the diameter of the nails being used. When nails split joist ends, they remain insecure. The little effort of drilling is essential. Specifications may require joists to be secured using straps or triple grips. After all joists are in place, solid blocking or herringbone

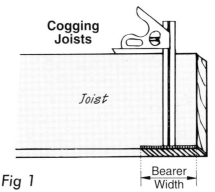

Fig 1

Set combination square to the depth of the narrowest joist and mark out the waste ready to remove with a chisel.

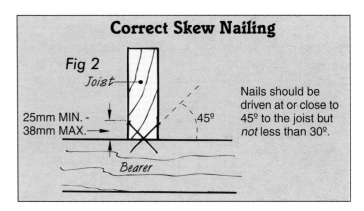

Correct Skew Nailing

Fig 2

Joist

25mm MIN. - 38mm MAX.

45º

Bearer

Nails should be driven at or close to 45º to the joist but *not* less than 30º.

strutting is installed *(if required)*. Where sheet flooring is to be applied all sheet join support blocking or noggings are installed.

Nailing Joists (nominal)

Joists are skew nailed to the plates and bearers with 2/75x3.05mm MIN. nails at a 45° angle to each crossing and *not* less than 30° (see fig 2). Use flathead for softwood. Additional nails may be required per joint. Check plans and specifications.

Solid Blocking & Herringbone Strutting

When the depth of joists is four times their breadth (thickness) or greater they should have either a continuous trimmer joist, solid blocking or herringbone strutting between the outer pairs of joists and at all intermediate support bearers and walls @ 1800mm ¢ MAX. In addition, where deep joists are unseasoned and where the span is over 3000mm and ceiling lining is *not* installed on the underside,

continuous blocking or herringbone strutting should be fitted between joists in rows 1800mm apart MAX.

Trimmers and solid blocking can be 25mm MIN. thick *(see herringbone strutting Page 68)*. Blocking and strutting should be fixed 12mm MIN. inside the upper and lower edges. However, when solid blocking is to be utilised for sheet edge support, it is kept flush with the top of joists.

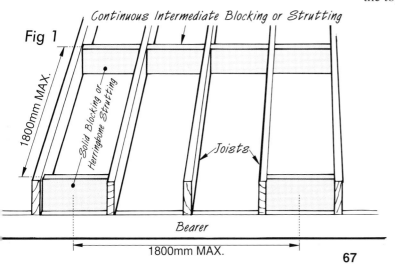

Fig 1

Continuous Intermediate Blocking or Strutting

1800mm MAX.

Solid Blocking or Herringbone Strutting

Joists

Bearer

1800mm MAX.

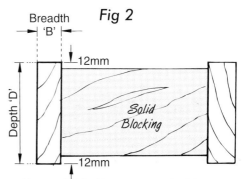

Fig 2

Breadth 'B'

12mm

Depth 'D'

Solid Blocking

12mm

Keep solid blocking 12mm under joist tops and bottoms to allow for shrinkage and air circulation unless its being utilised as sheet edge support see above text.

Herringbone Strutting

Herringbone strutting has in the main been replaced by solid blocking. However, a knowledge of how to cut such angles may be useful. These are usually cut out of 45x45mm, 45x35mm or 35x35mm timber.

How to Find the Length & Bevel of Herringbone Struts

Step 1 Herringbone struts are fixed 12mm down from the joist tops and 12mm up from the bottoms. To obtain their lengths and bevels, measure the joist depth and deduct the above 12+12=24mm. This will be the strutting depth. Mark this measurement on the tongue of a rafter square measuring from the external corner of the body. Then mark the spacing between the joists on the lower edge of the body.

Step 2 The square is then laid on a length of strutting material as in fig 1 and the bevel lines squared across. The strut can be cut to length.

If the joists have been laid to accurate spacings using a spacing nogging, then the first strut can be used as a pattern for cutting the remainder. Some spacings at ends will not be standard. These have to be marked out in the same manner and cut individually.

Securing the Struts

To fix them in alignment a chalk line in sprung through the top edges of the joists and a line is squared down the face of both sides of each joist. The struts are nailed to the lines 12mm clear of the joist tops and bottoms.

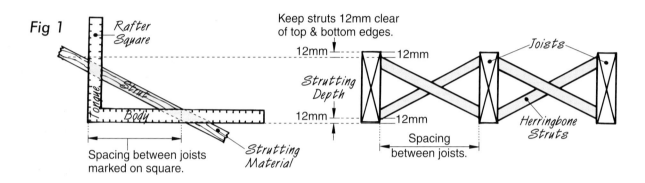

Fig 1

Preparing to Lay Flooring

Prior to laying flooring for either cut-in (fitted) flooring or platform flooring, the joist tops need to be all on the same level plane. Firstly, check the tops by laying a straightedge through as in fig 2. Dress any spring or bumps back with an electric plane. Ensure also that all sheet supporting noggings are flush, and that blocking or struts are below the top of the joists as recommended.

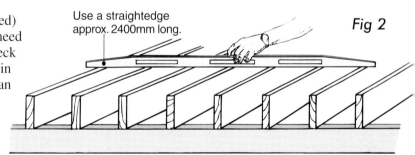

Use a straightedge approx. 2400mm long.

Fig 2

Laying Sheet Flooring
(such as Structaflor Particleboard)

Joists & Noggings for Sheet Flooring

Joists are spaced @ 450¢ for 19mm thick particleboard and 600¢ for 22mm thick particleboard.

Square Edge Sheets can be laid parallel or at right angles to joists. Sheet ends and sides require nogging, joist or trimmer support.

Tongue & Groove Sheets are laid at right angles to joists and *don't* require noggings to support the edges. Sheets must cover a minimum of two joist spacings. However, sheet ends do require support.

If using the platform method, when sheet flooring arrives on the job, have it off loaded directly onto the floor joists but not in one stack.

Prior to sheets arriving, carry out the following:
   a). Provide joists and/or nogging support to the sides and ends of panels as required.

   b). Check the joist surface for any high spots as above.

   c). Spring the chalk alignment line as in Step 2 and fig 1, Page 69.

FASTENING REQUIREMENTS (for Structaflor)

FASTENER TYPES	JOIST TYPE	FASTENER	STRUCTAFLOR/ TERMIFLOR	FASTENER (mm)
Manual Nailing	H.W.D. or Cypress	Bullet or flathead	19, 22mm	50x2.8
Manual Nailing	H.W.D. or Cypress	Bullet or flathead	25mm	65x3.75
Manual Nailing	Softwood	Bullet or flathead	19, 22mm	65x2.8
Manual Nailing	Softwood	Bullet or flathead	25mm	75x3.75
Machine Driven	H.W.D. or Cypress	D Head, round head or finished head	19, 22mm	50x2.5
Machine Driven	H.W.D. or Cypress	D Head, round head or finished head	25mm	65x2.5
Machine Driven	Softwood	D Head, round head or finished head	19, 22mm	65x2.5
Machine Driven	Softwood	D Head, round head or finished head	25mm	75x2.5
Screw Fixing	All Timbers	Type 17 countersink self-drilling screws	19, 22mm	No. 10x50
Screw Fixing	All Timbers	Type 17 countersink self-drilling screws	25mm	No. 14x65
Screw Fixing	Steel	Countersink, self embedding head, self drilling screws, preferably with self breaking cutter nibs	19, 22, 25mm	No. 10x45

Notes: a). *Use galvanised nails for exposed platform construction and designated wet areas.* **b).** *Screw bullet head nails for improved holding power.* **c).** *Steel screws should be suitably coated to resist corrosion.* **d).** *QuickDrive has a screw system that provides high pull-down and long term holding power when compared to nails and allows fastening from a standing position.*

Joist & Fastener Spacing
Joistes are spaced at 450¢ for 19mm thick particleboard & 600¢ for 22mm & 25mm thick particleboard.
Space Fastners at 150¢ to perimeter edges and 10mm from square edge and 25mm for T&G edges. Space them at 300mm ¢ In the body of the sheet and on all joins.

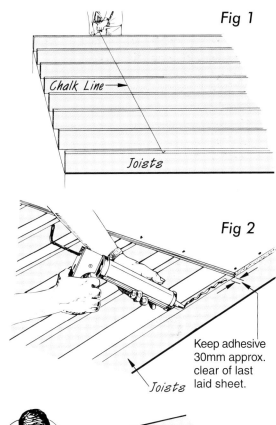

Fig 1

Chalk Line → Joists

Fig 2

Keep adhesive 30mm approx. clear of last laid sheet.

Joists

How to Lay Sheet Flooring (such as Structaflor)

Step 1 Check the joist tops are true and in alignment as in fig 2, Page 68.

Step 2 Spring a chalk line across the joist tops and through the length of the floor one sheet width in from one edge. Drive some 50mm nails into the joists to the line as a guide for positioning the sheets. Apply a 5mm bead of structural flooring grade adhesive to joist tops, enough for one sheet at a time. At joins provide a fresh bead for the adjoining sheet.

Step 3 Lay the sheets correct side up and in alignment with the chalk line. Lay before a skin forms on the adhesive. Skinned adhesive should be removed. For a fastening guide, spring a chalk line across the flooring indicating the centres of all joists. Nail or screw fasten from the sheet centres out towards the edges. **Use fasteners and spacings recommended by the manufacturer.**

Drive nails flush with the surface and just prior to sanding punch them 2mm below. Ensure all fastening is complete within 15 minutes of laying each sheet. Lay adjoining sheets in a staggered pattern as in fig 4. Take care that adhesive is *not* squeezed up onto sheet edges preventing tight joints. To avoid it is best to terminate adhesive beads approx. 30mm from joins. When laying T & G sheets, ensure all dust and chips have been cleared from the grooves and edges.

Step 4 On one side and edge, sheet excess may overhang the floor perimeter. To remove, spring a chalk line through and trim with a circular saw (see fig 3). Seal all raw edges with adhesive. Use a spatula or putty knife.

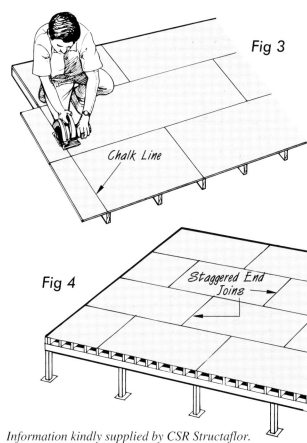

Fig 3

Chalk Line

Fig 4

Staggered End Joins →

Information kindly supplied by CSR Structaflor.

Note: a). *Don't* leave the floor surface exposed to weather for longer than recommended; and
b). *Don't* lay plastic sheeting or apply sealants over the exposed floor in order to protect the floor from moisture as these measures only trap moisture and retard drying.

T & G Flooring

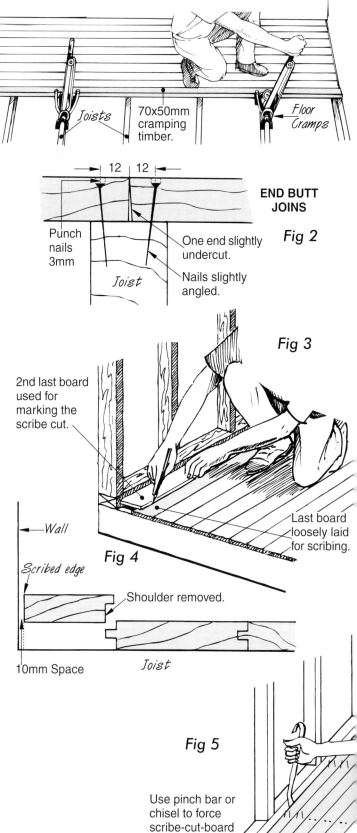

Fig 1

T & G flooring should be seasoned and have 10-12% moisture content on delivery. A certificate should be requested stating the average moisture content. Flooring with a greater moisture content can be left loosely laid in place for approx. two weeks to help reduce the moisture content.

T & G *should not* be laid as a platform floor but rather should be installed after the house roof and walls are clad and windows installed. Alternatively, a platform floor can be laid using particle board flooring and 12mm thick T & G flooring laid over the surface after enclosing the building.

Subfloor Ventilation — *See Page 25*. Subfloor ventilation is essential to ensure a stable floor.

Joists — Should be the thickness and spacing stipulated in approved plans. When designing joist spacing, consideration must also be given to the flooring, species, thickness, grading and whether it is to be butt jointed or end matched. *For full details see AS 1684.2*

Laying T & G Flooring

Step 1 Allow a 10mm gap between the first board and the wall and fasten the first row of boards to a chalk line driving the nails fully.

Step 2 Lay out four or five rows of boards against the first board and apply floor cramps. Fit a length of straight 75x50mm timber between the floor cramps and T & G boards to prevent bruising (fig 1). Make sure the boards are tight before nailing.

When cramping, move the cramps to a different pair of joists each time. This contributes to keeping the boards straight. End joins on end matched boards can fall between joists. Butt joins should be cut to fall centrally on joists. All joins should be staggered. Slightly undercut one side of butt joins to ensure a tight fit (fig 2). All nails should be driven at a slight angle to give increased holding power. On cut-in flooring allow a 10mm space between flooring ends and the wall framing.

Step 3 When laying the last board on a cut-in floor, leave a 10mm gap between the board and the wall plate and scribe the board to be cut as in figs 3 & 4. Tighten the last board using a pinch bar. Partially drive the nails into the board. Then lever the board tightly into place. While maintaining pressure on the pinch bar, drive the nails home (see fig 5).

Surface Finishing

Many commonly applied surface finishing materials including some polyurethanes can prevent the normal expansion and contraction of individual boards resulting in uneven or unsightly cracking.

For this reason, it is recommended to use proven flooring grade Tung oil with resin preparations. *Not polyurethanes.*

Expansion Joints

Provide a 10mm expansion gap to fall under partitions for continuous floors over 6m measured at right angles to the flooring.

Fastening Requirements: For 19-20mm thick boards — 50x2.8mm bullet head MIN. *(hand driven)* or 50x2.5mm MIN. *(machine driven)* into HWD or cypress joists; and 65x2.8mm MIN. bullet head *(hand driven)* or 65x2.5mm MIN. *(machine driven)* into softwood. Fasten all boards with two nails per joist.

Installing Scyon Secura™ Interior Flooring

(Scyon is an alternative flooring to compressed sheet; is lighter and gun nailable).
(The following details are for tiled internal wet areas)

Preparing to Lay

All joists must be seasoned and be 45mm thick at end joins to enable fastening. Space at 450mm centres. Joists and framing must be straight, level and the tops in alignment with one another as in Page 68, fig 2, and comply with AS 1684.2.

Joining to Particleboard or Plywood

Joins can be made on a joist or trimmer (see fig 1) or off the joists as in fig 2.

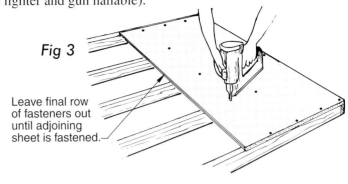

Fig 3

Leave final row of fasteners out until adjoining sheet is fastened.

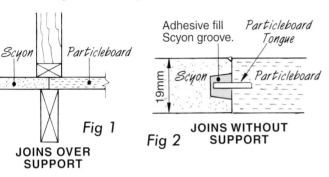

Fig 1 — JOINS OVER SUPPORT

Adhesive fill Scyon groove. Particleboard Tongue

Particleboard

Scyon

19mm

Fig 2 — JOINS WITHOUT SUPPORT

Fastenings — *Timber Joists:* 2.8x50mm FC nails or 50mm 'D' head gun nails or No.8x40mm CSK self drilling screws; *Steel Joists:* 0.75-1.6mm BMT use 40mm Hardi Drive screws; over 1.6mm BMT use 40mm CSK self drilling screws. *(See fig 4 for nailing pattern and refer to fastener manuf. for more information).*

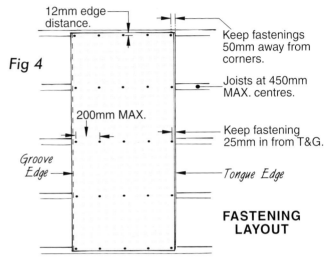

12mm edge distance.

Fig 4

200mm MAX.

Groove Edge

Keep fastenings 50mm away from corners.

Joists at 450mm MAX. centres.

Keep fastening 25mm in from T&G.

Tongue Edge

FASTENING LAYOUT

How to Lay

Sheets are laid at right angles to joists in a staggered pattern as in fig 4, Page 69. This may necessitate cutting an adjoining sheet as in fig 7.

Step 1 Lay one continuous bead of suitable construction adhesive to all joists within the area of the sheet being laid.

Step 2 Lay the first sheet and drive fastenings accord. to the pattern as in fig 4. Fastenings at joins should be left out until after fastening adjacent sheets (see fig 7).

Step 3 Install the second sheet, bring the tongue and groove joint together and fasten in place. Where a butt joint occurs drive two 2mm dia. nails as spacers to create a gap to receive sealant later (see fig 7). After fixing adjoining sheets, remove spacer nails.

Step 4 Fill tongue and groove, butt joins and nail holes with James Hardie Joint Sealant (see fig 8).

Step 5 Clean the surface with a damp cloth and lay tiles or other accord. to manuf. recommendations.

> *Note:* Tiling should be carried out within 3 months.

> *For further details including safety instructions refer to the James Hardie Product installation instructions.*

Edge distance 12mm

2mm gap filled and spread over butt joint with James Hardie Joint Sealant.

2 continuous beads of construction adhesive.

Nails as in text.

Fig 5 — Joist

BUTT JOINS

James Hardie Joint Sealant

25mm

Edge distance

Scyon

Fig 6

Fastenings

T&G JOINS

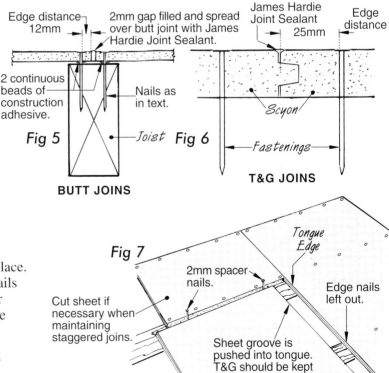

Fig 7

Tongue Edge

2mm spacer nails.

Edge nails left out.

Cut sheet if necessary when maintaining staggered joins.

Sheet groove is pushed into tongue. T&G should be kept free of adhesive.

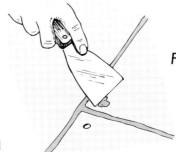

Fig 8

Fill joins and fastening heads with James Hardies Joint Sealant and spread with spatula to form a seal between sheets.

4

Wall Framing

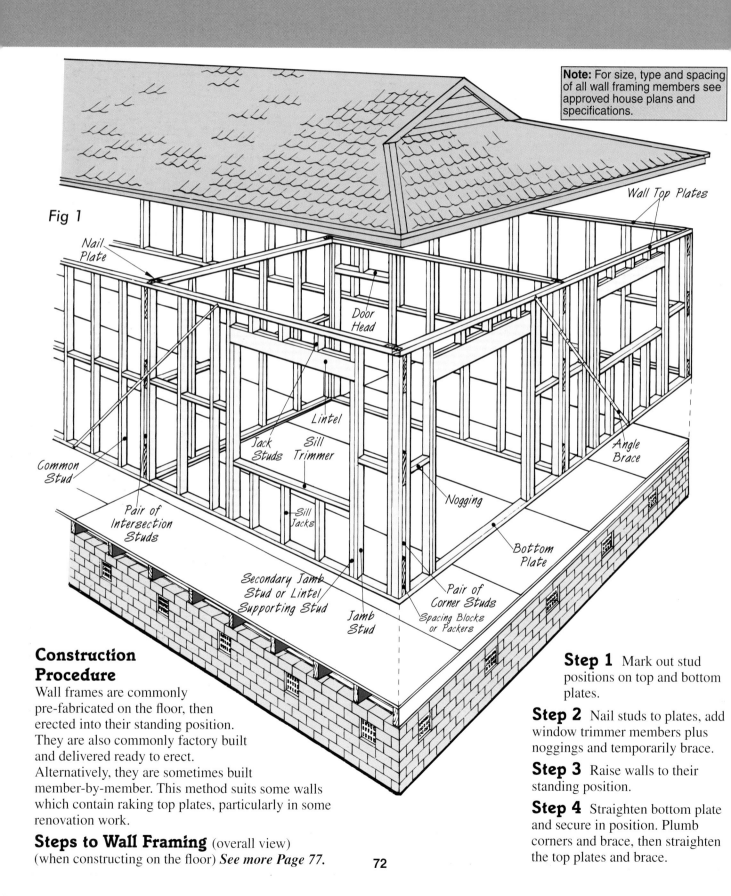

Fig 1

Note: For size, type and spacing of all wall framing members see approved house plans and specifications.

Wall Top Plates

Nail Plate

Door Head

Lintel

Jack Studs

Sill Trimmer

Common Stud

Pair of Intersection Studs

Sill Jacks

Secondary Jamb Stud or Lintel Supporting Stud

Jamb Stud

Nogging

Pair of Corner Studs

Spacing Blocks or Packers

Bottom Plate

Angle Brace

Construction Procedure

Wall frames are commonly pre-fabricated on the floor, then erected into their standing position. They are also commonly factory built and delivered ready to erect. Alternatively, they are sometimes built member-by-member. This method suits some walls which contain raking top plates, particularly in some renovation work.

Steps to Wall Framing (overall view)
(when constructing on the floor) *See more Page 77.*

Step 1 Mark out stud positions on top and bottom plates.

Step 2 Nail studs to plates, add window trimmer members plus noggings and temporarily brace.

Step 3 Raise walls to their standing position.

Step 4 Straighten bottom plate and secure in position. Plumb corners and brace, then straighten the top plates and brace.

Wall Plates

Wall plates provide a means of installing and erecting studs, and for attaching the upper and lower edges of wall linings. The bottom plates are utilised for fastening the wall to the floor and the top plates for supporting the roof framing. In some instances a second top plate is specified. These are known as ribbon plates or double top plates. The upper plate can sometimes be increased in width to double as ceiling battening (see Page 123).

Joining Wall Plates to Beams
See Pages 74&75 for details.

Joining Wall Plates End-to-End
Top plates may be joined end-to-end as illustrated below. Ensure these joins *don't* arrive above openings. End-to-end joins in top plates should be joined as in figs 3 to 8. Use the method which has been specified in approved plans. Bottom plates can be butt joined over solid support.

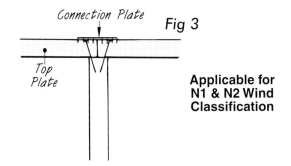

Connection Plate

Fig 3

Top Plate

Applicable for N1 & N2 Wind Classification

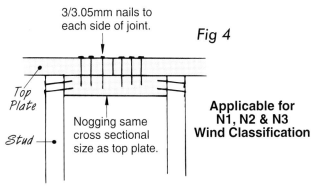

3/3.05mm nails to each side of joint.

Fig 4

Top Plate

Stud

Nogging same cross sectional size as top plate.

Applicable for N1, N2 & N3 Wind Classification

END-TO-END JOINS IN TOP PLATES

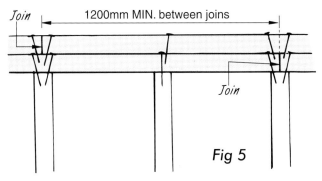

Join

1200mm MIN. between joins

Join

Fig 5

Applicable for N1, N2 & N3 Wind Classification

JOINS IN RIBBON (DOUBLE) TOP PLATES   73

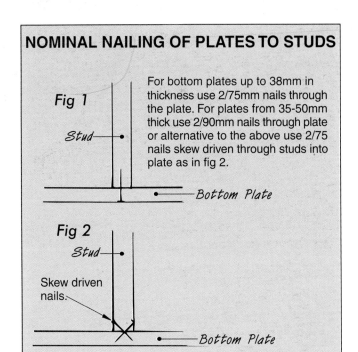

NOMINAL NAILING OF PLATES TO STUDS

Fig 1

Stud

For bottom plates up to 38mm in thickness use 2/75mm nails through the plate. For plates from 35-50mm thick use 2/90mm nails through plate or alternative to the above use 2/75 nails skew driven through studs into plate as in fig 2.

Bottom Plate

Fig 2

Stud

Skew driven nails.

Bottom Plate

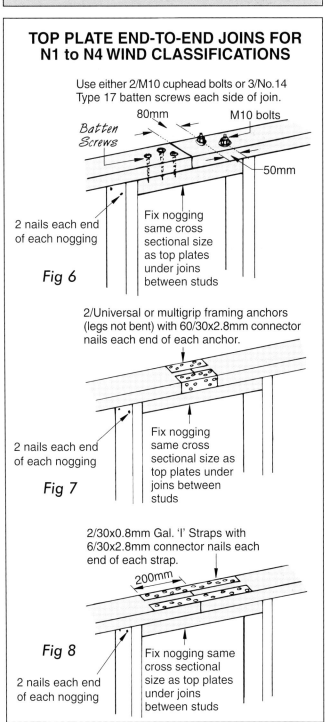

TOP PLATE END-TO-END JOINS FOR N1 to N4 WIND CLASSIFICATIONS

Use either 2/M10 cuphead bolts or 3/No.14 Type 17 batten screws each side of join.

Batten Screws

80mm

M10 bolts

50mm

2 nails each end of each nogging

Fix nogging same cross sectional size as top plates under joins between studs

Fig 6

2/Universal or multigrip framing anchors (legs not bent) with 60/30x2.8mm connector nails each end of each anchor.

2 nails each end of each nogging

Fix nogging same cross sectional size as top plates under joins between studs

Fig 7

2/30x0.8mm Gal. 'I' Straps with 6/30x2.8mm connector nails each end of each strap.

200mm

Fig 8

2 nails each end of each nogging

Fix nogging same cross sectional size as top plates under joins between studs

Cont.

Joining Wall Top Plates

Join top plates at corners and intersections with connector plates using 3/35x3.15mm gal. connector nails each side of the joint.

Beams to Wall Plate Connection

See figs 3&4.

The top of the beam is usually cut to fit around the top plate as in fig 4.

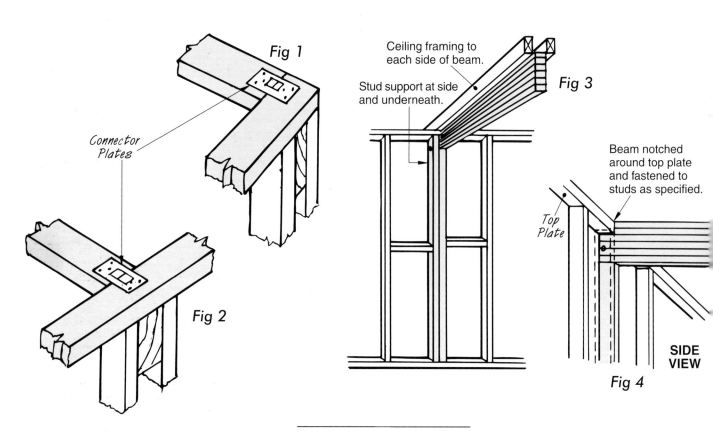

Fig 1

Connector Plates

Fig 2

Ceiling framing to each side of beam.

Stud support at side and underneath.

Fig 3

Beam notched around top plate and fastened to studs as specified.

Top Plate

SIDE VIEW

Fig 4

Attaching Internal Bracing Walls to External Walls (fig 5)

Internal bracing walls should be attached to the roof framing and/or external walls. Depending on the design strength required, this may require strapping to external walls as in fig 5 as well as other connections. *Refer to approved plans and specifications.*

Verandah Beam Plates to Post Joint

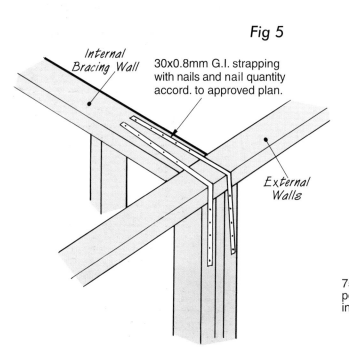

Fig 5

Internal Bracing Wall

30x0.8mm G.I. strapping with nails and nail quantity accord. to approved plan.

External Walls

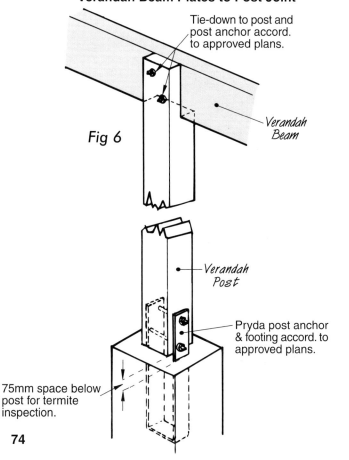

Fig 6

Tie-down to post and post anchor accord. to approved plans.

Verandah Beam

Verandah Post

Pryda post anchor & footing accord. to approved plans.

75mm space below post for termite inspection.

Cont.

Cantilevered Beam & Top Plate Connection

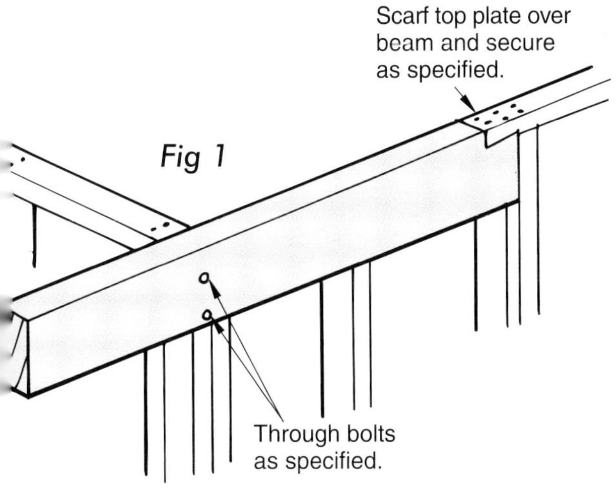

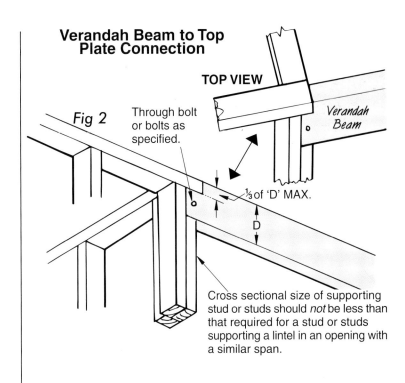

Scarf top plate over beam and secure as specified.

Fig 1

Through bolts as specified.

Verandah Beam to Top Plate Connection

TOP VIEW

Fig 2

Through bolt or bolts as specified.

Verandah Beam

⅓ of 'D' MAX.

D

Cross sectional size of supporting stud or studs should *not* be less than that required for a stud or studs supporting a lintel in an opening with a similar span.

Localised Stiffening of Plates

At locations where the **plates** are required to carry a concentration of loading from: a roof beam, roof strut, girder truss; large floor, roof or ceiling loads, stiffening is provided in the form of a stud directly below as in fig 3. Dimensions of this stud should be designed accord. to 'AS 1684.2'. Alternatively, where specified, use solid blocking as in fig 4.

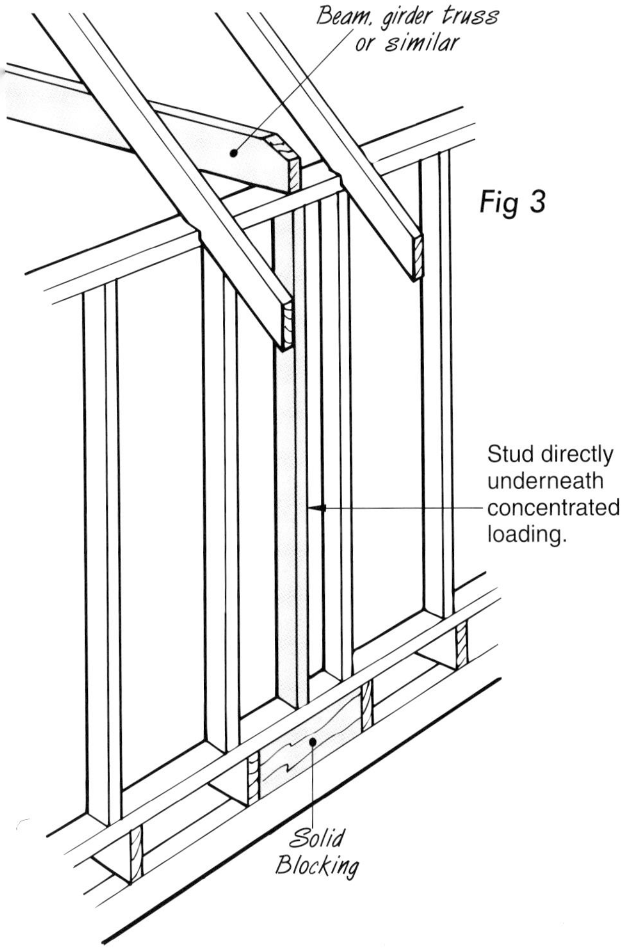

In openings which are greater than 1200mm wide the bottom plate should be stiffened directly below the jamb studs with either a slab floor, a joist or solid blocking as in fig 5.

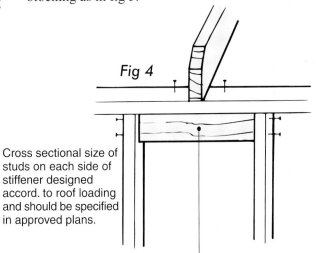

Fig 4

Cross sectional size of studs on each side of stiffener designed accord. to roof loading and should be specified in approved plans.

Stiffener blocking is the same size as the adjoining studs placed on edge with 2/3.15 nails each end and 2 in the top edge. When area supported exceeds 10m² but is *not* more than 15m² use two stiffener blocks with 3 nails each end.

Beam, girder truss or similar

Fig 3

Stud directly underneath concentrated loading.

Solid Blocking

Fig 5

Lintel

Opening over 1200 Wide

Slab floor, joist or blocking directly underneath jamb studs.

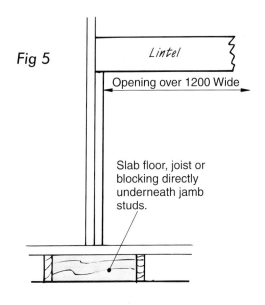

Stud & Jack Stud Lengths

Stud Size & Spacing

The size and spacing of studs is determined by the load they will carry and the wall linings they will be supporting. These dimensions are shown in approved plans and specifications.

Preparing Studs

Studs are squared and trimmed to their respective lengths. Keep straight and sprung studs in separate stacks arranged closely and neat but away from the sun and wind if possible. Use as soon as possible after cutting.

When the standard studs are all cut, continue cutting the remaining trimmer studs, then sill and jack studs. Lastly, cut the standard noggings: then the smaller odd shaped noggings. Notice the procedure is to cut from the longest stud component down to the smallest nogging. This method utilises offcuts. Badly sprung lengths can be cut into shorter sizes. Keep straight timber for trimmer studs, window sills and for walls which require a straight surface such as for panelling and ceramic tiling. However, studs can be straightened see Page 90.

Common Stud Length

Following is an example of how to arrive at the correct common stud length. These measurements will vary with the various material sizes used. However, the basic method of arriving at the stud length remains the same.

Example to find the stud length for typical situation as in fig 1.

Take ceiling height		*2410mm*
Add *ceiling lining thickness*	*+ 12mm*	
		2422mm
Deduct *top & bottom plates*	*- 80mm*	
(thickness of 35mm for		
bottom & 45mm for top)		*2342mm*
Add *10mm as a clearance*	*+ 10mm*	
(for fitting wall sheets)		
Stud Length	**=**	**2352mm**

The Storey Rod

For cutting the remaining framing members, make a storey rod to establish trimmer, lintel and sill jack stud lengths. Take a straight length of 35x35mm and mark plates and all lintel and sill positions accurately. Write on each their respective room positions. One or more sides of the rod could be used. Keep one side for brick coursing on brick veneer houses.

The underside of lintels as a rule, continue at the same height throughout. Usually if sliding aluminium patio doors are used, this lintel height will become the governing height for all lintels. Add a 12mm gap between joinery heads and timber lintels for possible lintel deflection. Mark off the bottom and top sides of each lintel. The space

remaining above the lintel is the *lintel jack stud length*. The space below the lintel to the bottom plate is the *jamb or trimmer stud length*.

Then, establish sill heights by measuring down from the underside of the lintels plus the 12mm deflection allowance and a further 3mm allowance for levelling up of windows later. Mark top and bottom sides of sill trimmers. The remaining space below the sill will be *sill jack lengths*. It is advisable to establish opening sizes by physically measuring the actual window frame.

For Brick Veneer:

One side of the storey rod should contain both the vertical and horizontal brick joint spacings. *For brick heights and lengths (see Page 12).* The bricklayer should be made aware of your layout planning.

Brick stretchers (horizontal courses) terminate at openings in whole or half brick sizes. Use the rod to establish where bricks will terminate beside openings and adjust the jamb studs accordingly. Heights of sills and lintels allow more flexibility as the sill brick can have its angle of slope adjusted or the course directly below cut to a reduced thickness.

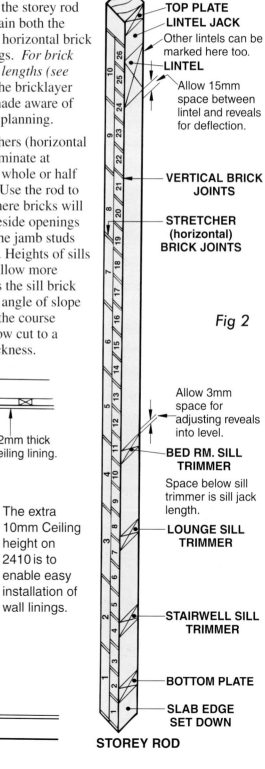

Fig 2

Fig 1

Stud

12mm thick ceiling lining.

Stud Length 2352mm

Ceiling Height 2410mm

The extra 10mm Ceiling height on 2410 is to enable easy installation of wall linings.

45

35

STOREY ROD

TOP PLATE
LINTEL JACK
Other lintels can be marked here too.
LINTEL
Allow 15mm space between lintel and reveals for deflection.

VERTICAL BRICK JOINTS

STRETCHER (horizontal) BRICK JOINTS

Allow 3mm space for adjusting reveals into level.

BED RM. SILL TRIMMER
Space below sill trimmer is sill jack length.

LOUNGE SILL TRIMMER

STAIRWELL SILL TRIMMER

BOTTOM PLATE

SLAB EDGE SET DOWN

Stud Positioning

Marking Out Wall Plates

Set out long lengths for top and bottom plates and cut to length. Ensure both ends are square and tack them together. Lay them in their proposed position on their edges on the floor. This procedure gives a check for correct length and enables speedy location of plates rather than storing them in a stack. Keep the sprung edges on the same side.

Stud Spacing — Stud size and spacing are governed by design factors at the pre-planning stage and will be specified in the approved plans and specifications.

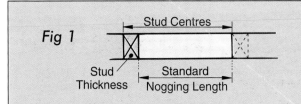

Fig 1

Standard Nogging Length

To determine standard nogging length, mark stud centres on a length of plate material. Then mark off one stud thickness from one centre line. The remaining space to the opposite centre line will be the standard nogging length. Use this first nogging as a pattern for marking off stud positions and for cutting the remaining standard noggings. Make the pattern nogging the same thickness as the studs in order to be able to mark stud thicknesses on the plates as in fig 2.

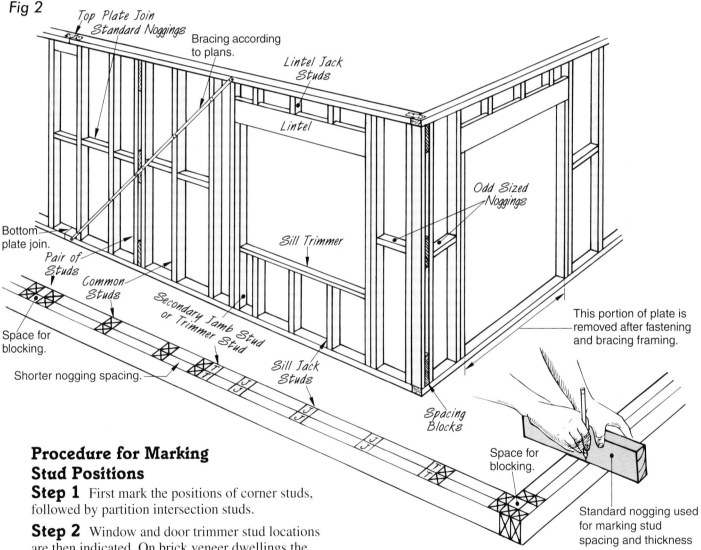

Fig 2

Procedure for Marking Stud Positions

Step 1 First mark the positions of corner studs, followed by partition intersection studs.

Step 2 Window and door trimmer stud locations are then indicated. On brick veneer dwellings the sides of proposed window frames in openings should coincide with vertical brick joints. If there are any studs supporting exposed beams, mark these positions as well.

Step 3 Some wall linings require a stud to support a vertical sheet join. Position these at this point. Then using the standard nogging as a gauge block, mark off all the remaining common stud positions. This is done in the same manner as described for indicating joist positions as in figs 2 & 3, Page 65.

Whenever one of the previously marked studs positions is arrived at, there will be an odd spacing smaller than the standard nogging. It is standard practice to disregard this and not attempt to equalise the stud positions. Shorter noggings are later cut to fit all these odd spaces.

Step 4 Roof trusses and ceiling joist positions can also be marked at this stage on the top side of the top plate. However, this could be postponed until after the frame is erected.

Corner & Pairs of Studs

Pairs of Studs

Pairs of studs are required at corners and 'T' junctions as in figs 1 to 6. These are spaced apart with blocks from the same material as the studs x 200mm long. Blocks are spaced at 900mm ¢ vertically MAX.

For brick veneer, when ply bracing is to be fastened to the outer intersecting corner, use figs 1&2, otherwise figs 3&4 can be applied to save one stud.

For pairs of studs at intersections, blocks are used on their flat as in figs 5 & 6.

Spacing-blocks are easier to install after the frames have been erected, plumbed and braced. Install them before installing the short noggings which have been left out at wall ends.

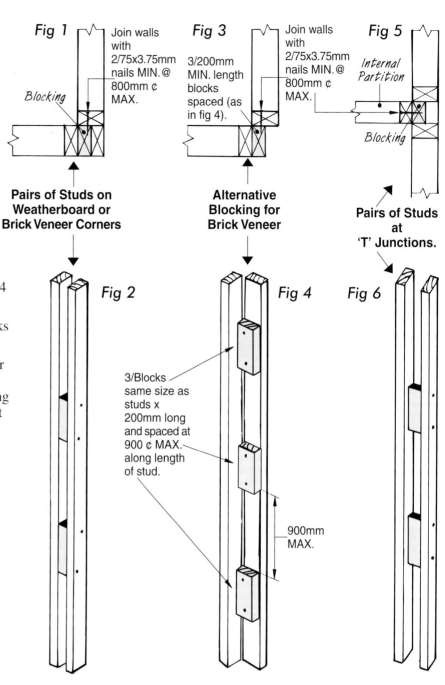

Fig 1 — Join walls with 2/75x3.75mm nails MIN. @ 800mm ¢ MAX. Blocking

Pairs of Studs on Weatherboard or Brick Veneer Corners

Fig 3 — 3/200mm MIN. length blocks spaced (as in fig 4).

Alternative Blocking for Brick Veneer

Fig 5 — Internal Partition. Join walls with 2/75x3.75mm nails MIN. @ 800mm ¢ MAX. Blocking

Pairs of Studs at 'T' Junctions.

Fig 2

Fig 4 — 3/Blocks same size as studs x 200mm long and spaced at 900 ¢ MAX. along length of stud.

900mm MAX.

Fig 6

Studs at Sides of Openings

In openings up to 900mm wide and where door jambs are to be attached to jamb studs, and also where the studs will *not* be supporting concentrated loads, a single stud of the same depth as the common studs can be used at the sides of openings.

In openings where jamb studs are required to be of greater width the jamb stud is commonly built up of two or more studs. Secondary jamb studs form part of the total number of studs required at sides of openings. Those directly underneath the lintel are known as secondary jamb studs. The jamb stud or studs are fastened to the bottom and top plates. The lintel ends butt into the jamb studs. Lintels are end nailed in place through the jamb studs as in fig 1 and rest on the secondary jamb stud or studs.

Joining Jamb Studs
Join jamb studs together with 1/75mm nail @ 600mm ¢ for studs up to 38mm thick & for studs up to 50mm thick MAX. use 1/90mm nail @ 600mm ¢. Use 3.05mm dia. MIN.

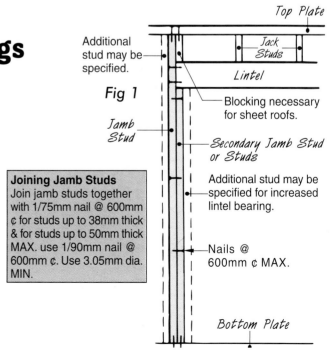

Fig 1

Top Plate

Additional stud may be specified.

Jack Studs

Lintel

Blocking necessary for sheet roofs.

Jamb Stud

Secondary Jamb Stud or Studs

Additional stud may be specified for increased lintel bearing.

Nails @ 600mm ¢ MAX.

Bottom Plate

Cutting Studs

Studs are best cut to length with a drop saw or radial arm saw. Fasten a length-stop-block onto the bench. Be sure to clear away sawdust build-up beside the block frequently. When these saws are unavailable the following method can be used.

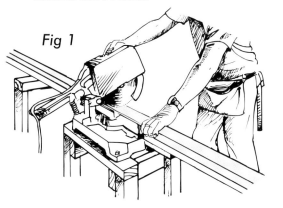

Fig 1

Notes:
a). Measure twice, cut once.
b). The length of studs in internal walls will be greater where ribbon top plates are used in external walls.

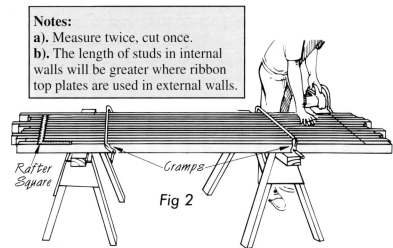

Rafter Square

Cramps

Fig 2

Lay out enough studs on stools to fit the largest cramps available. Studs can be laid on their edges or sides depending on the circular saw blade depth. Keep all springs uppermost and top edges flush, square the ends to stud length using a rafter square. Diagonal measurements can be made for greater accuracy. Then trim studs to length using a circular saw as in fig 2.

Lintels

Lintels support the top plate, ceiling and roof above openings. It is advisable that any lintels over 175mm in depth be seasoned.

Various methods of construction are possible. Each one satisfies a specific design requirement. The approved plans should indicate the specific method to use. Lintels housed into studs (see fig 1) can only be used for openings up to 900mm wide. The housing depths may only be one quarter of the studs thickness or 10mm whichever is less. For all others it is common practice to use secondary jamb studs to rest the lintel on. The approved plans will state the size and number of jamb and secondary jamb studs required.

Concentrated Loads above Lintels

Concentrated loads above lintels should be avoided if possible. However where this is unavoidable. Approved plans should specify appropriate lintel designs.

Fig 1

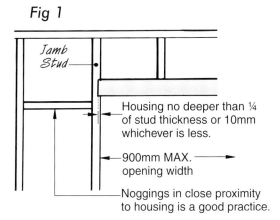

Jamb Stud

Housing no deeper than ¼ of stud thickness or 10mm whichever is less.

900mm MAX. opening width

Noggings in close proximity to housing is a good practice.

Install blocking at each end for sheet roofs.

Solid One Piece Lintel

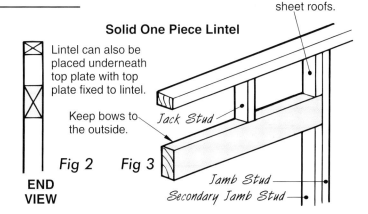

Lintel can also be placed underneath top plate with top plate fixed to lintel.

Keep bows to the outside.

Jack Stud

Fig 2
Fig 3

END VIEW

Jamb Stud
Secondary Jamb Stud

Lintel at Head Height

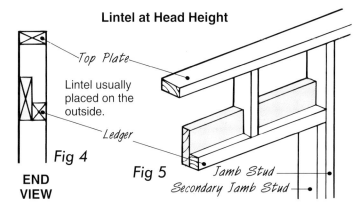

Top Plate

Lintel usually placed on the outside.

Ledger

Fig 4

Fig 5

END VIEW

Jamb Stud
Secondary Jamb Stud

Lintel Directly Below Top Plate

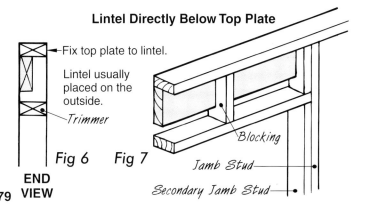

Fix top plate to lintel.

Lintel usually placed on the outside.

Trimmer

Fig 6
Fig 7

Blocking

Jamb Stud

Secondary Jamb Stud

END VIEW

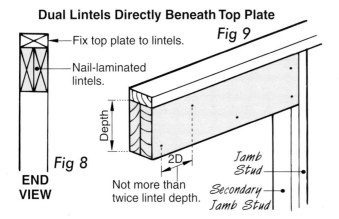

Dual Lintels Directly Beneath Top Plate

Fig 9

- Fix top plate to lintels.
- Nail-laminated lintels.

Depth

2D

Not more than twice lintel depth.

Fig 8

END VIEW

Jamb Stud

Secondary Jamb Stud

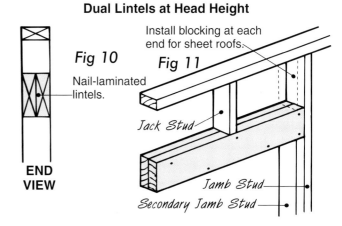

Dual Lintels at Head Height

Install blocking at each end for sheet roofs.

Fig 10

Fig 11

Nail-laminated lintels.

Jack Stud

Jamb Stud

Secondary Jamb Stud

END VIEW

Nail Laminating Lintels

Both lintels should be of the same species and stress grading. Nails should be 3.05mm dia. MIN. and should be spaced apart *not* more than twice the lintel depth and staggered as in figs 9 & 11. Nails should be through driven and clenched over or driven from both sides.

Wall Frame Assembly

Fig 1

How to Assemble

Step 1 Lay out plates on the floor in line with a chalk line as a guide for keeping the frame approximately square. Lay the bottom plate in proximity to its proposed erected position to avoid unnecessary handling of the assembled frame later.

Step 2 Install studs at each end of the frame first, then any trimmer and intersection studs as in fig 1.

Step 3 Fit lintels, sills and jack studs.

Step 4 Install remaining common studs.

Tops are square cut

Common Stud Position

Jack Stud Position

Lintel Positions

Secondary Jamb Studs

Jamb Studs

Pair of Intersection Studs

Proposed Opening

Chalk line on floor.

Sill Jack Positions

NAILING
Studs to Plates
2/75mm nails skewed through studs into plates or 2/75mm nails through 35mm thick plates into stud ends or 2/90mm nails through plates into stud ends in plates up to 50mm thick MAX. Use 3.05mm dia. F.H. nails for softwood.

Step 5 Install noggings as below.

Step 6 Square frame and fit bracing as in Pages 81-85.

Sills, Sill & Head Jacks & Nogging

Sills

Sills are usually cut out of common stud material. They are butt joined to the secondary jamb studs and fixed with two nails at each end.

Sill & Head Jacks

Sill and head jacks can be spaced at the common stud spacings or they can be equally spaced but *not* further apart than common stud spacings. To obtain

NAILING
Sill & Jacks
Use 75mm nails either skewed or through nailed. Use 3.05mm dia. F.H. nails for softwood and 3.05mm for hardwood and Cypress.

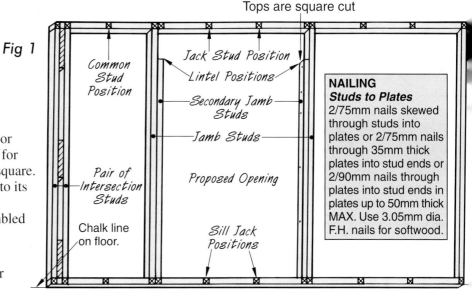

Fig 2

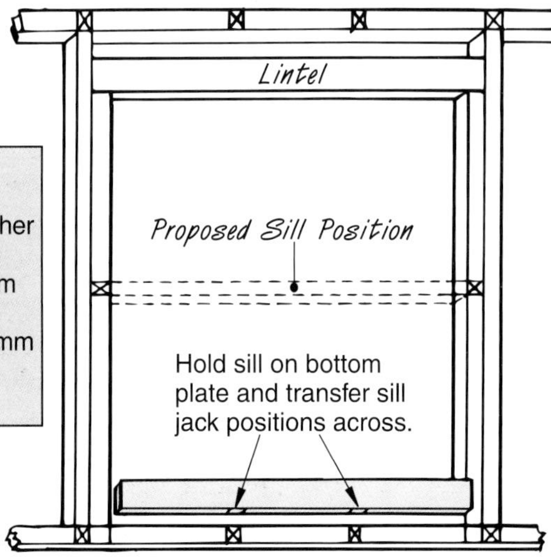

Lintel

Proposed Sill Position

Hold sill on bottom plate and transfer sill jack positions across.

jack positions on sills, hold sill against the bottom plate jack stud positions and simply transfer marks across (see fig 2). Fasten sill and head jacks with two nails at each end.

Noggings

One row of noggings is required at 1350mm ¢ MAX. to reduce stud twisting and to stiffen the centre of frames. In some cases, this will also provide a fixing base for wall lining joins.

Nogging may be installed on flat or on edge. If vertical board lining is being applied, attach rows of noggings as directed in manufacturer's specifications. Nogging is usually cut from the same stock as common studs and is best to be *not* less than 35mm thick. Noggings, when fixed in a staggered pattern, are easier to nail in place but *should not* be offset more than 150mm about the centre line. When they are used for joining linings, they should be 45mm thick to provide sufficient edge distance for nailing and are kept in line.

If Gyprock plasterboard sheets are laid horizontally,

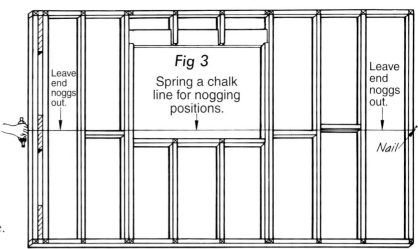

Fig 3
Spring a chalk line for nogging positions.

Leave end noggs out.

Leave end noggs out.

Nail

noggings must be kept away from the proposed horizontal joint.

How to Install Noggings

Spring a chalk line across studs as a nailing guide. Use four nails to each nogging, two at each end. Leave end noggings out of the frame until walls have been erected, plumbed and straightened. Leave the upper nails out of noggings where it is suspected housed bracing may cross. This will save damage to saw teeth when cutting slots or rebates for braces.

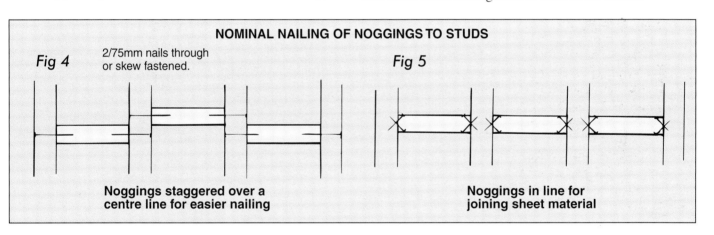

NOMINAL NAILING OF NOGGINGS TO STUDS

Fig 4
2/75mm nails through or skew fastened.

Noggings staggered over a centre line for easier nailing

Fig 5

Noggings in line for joining sheet material

Bracing Wall Frames

(much of the technical data from Pages 81-85 was kindly provided by Pryda Australia)

Temporary Bracing

The BCA requires that once walls are erected they must be braced. Temporary or permanent braces can be installed before erecting the frames. Permanent braces are usually attached initially with nails partially driven to enable fine adjustment later. Prior to their complete attachment, temporary braces equivalent to 60% of the strength of the permanent braces must be installed.

Purpose of Permanent Bracing

Permanent bracing is required at specified locations across the length and width of the house frame to resist wall racking and distortion of the frame through wind loads.

Bracing is engineered to the precise requirement therefore it is vitally important that care is taken to

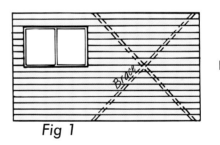

Fig 1

BRACED WALL

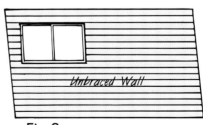

Fig 2

UNBRACED WALL
Exaggerated state of a wall without sufficient bracing. A wall that has racked only slightly out of alignment will cause wall linings and paintwork to crack.

install each one accord. to code and building manufacturer's requirements. Bracing type, size, locations and fastenings should be indicated on the approved plans,

Cont.

on the elevations, on the floor plan, or on a separate bracing plan.

It is important that designers clearly detail the bracing and fastening requirements for each bracing unit. The integrity of the building depends on adherence to these design requirements and *should not* be altered without consultation with the designer and building certifier.

How Wall Bracing Works (including Metal, Timber & Plywood)

Bracing is part of a *system* which incorporates elements from the roof down to the foundations. This system is designed to resist forces on the structure such as wind.

These forces may cause the building to rack sideways as in figs 1 & 2 or worse, cause total collapse, *bracing* is the main element providing resistance to racking. High winds may cause uplift of the structure. In this case the *holding-down* elements provide the main resistance.

The Wall Bracing system comprises:
a). The anchorage or tie-down — from the roof to the wall.

b). The bracing — the fixing of the bracing to the frame especially its end fixings including any straps which are part of the bracing unit.

c). The frame — to which the bracing is fixed i.e. studs, wall plates including any joins on those plates.

d). The connection — of the bracing unit to the floor or supporting structure.

e). The anchorage or tie-down — of wall-to-floor and *tie-down* of floor-to-foundations.

If one element is *not* properly attached it weakens the system. The system must be able to resist wind on the structure from all sides and above, therefore bracing must be applied both along its length and across its width. Bracing in the internal walls must resist wind load from the roof and on the floor. For this reason internal bracing units must be connected to the roof.

Loads in 2-Storeys

The resistance required for the lower storey of a 2-storey structure is greater than that required for the upper storey or in a single storey. Usually the higher the building the greater the wind loading upon that building.
The speed of wind and its force increases with height above ground i.e. the wind force at 10metres high is rated at 18% greater than at 4metres. Also, bracing in the lower storey must resist wind force on both storeys. For both of these reasons, bracing in

the lower of 2-storeys is required to be about 60% stronger than the upper storey or for single storey.

Types of Bracing

Bracing is chosen to suit the individual specification and application.

There are several types of bracing. Ensure the designer stipulates on the plan which type of braces are to be used. Incorrect application could lead to litigation in the event of failure.

Where cutting-in of braces is permitted, approved metal angle or timber bracing may be used. In some applications, the stud is *not* permitted to be weakened by the cutting in of bracing so speedbrace, flat strap or structural sheet ply or hardboard is used. Sheet bracing is also applied on short walls where a diagonal brace would be too steep to be effective.

> **Note:** *The following metal brace details are based upon engineered Pryda braces fastened with 35x3.15mm galvanised, Pryda timber connector nails.*

Angle Bracing (see figs 3-6)

These are only suitable for 'Type A' bracing units. Angle Brace (also known as 'Mini Brace') is only suitable for Type A bracing units and there are two alternative installations as shown on the following page, i.e 'One Length' type with stud ties or strap nails at three corners, or the 'Two Lengths' type with two diagonally opposed braces in the same wall. These braces must *not* be crossed over, cut, bent or flattened.

They are housed into the wall frames in a slot which is made by a single pass of the circular saw. For this reason some framing nails may need to be left out until the slot is made. Cut-in bracing reduces the studs strength. For this reason the Designer may specify stronger studs.

Angle bracing should be installed in a straight line and with the vertical leg downwards. Ends should *not* be further back from the sawn ends of plates than 150mm MIN. They should be fixed at the angles given in figs 3 or 5. The saw cut housing *should not* be cut deeper than 20mm. When the slot for Mini Brace is *not* deeper than the 16mm, the studs can have a 3mm rebate to permit the brace to be housed flush with the studs to permit claddings to pass across and remain flat.

Fastenings for 'Type A & B' Bracing

Bracing Unit	Fixing at Each Stud	Fixing at Wall Plates (Ends)
Type A	1/35x3.15mm Pryda nail	3/35x3.15mm Pryda nails
Type B	2/35x3.15mm Pryda nail	4/35x3.15mm Pryda nails

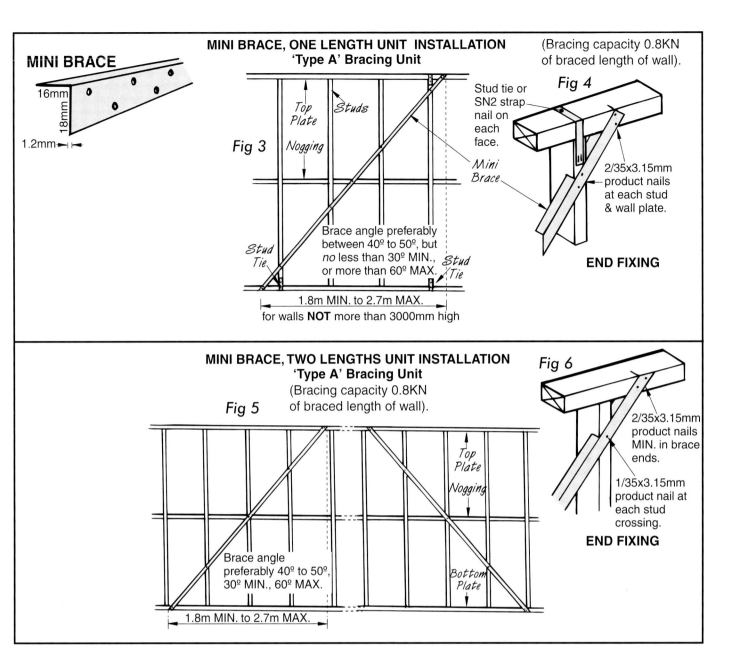

MINI BRACE

16mm
18mm
1.2mm

MINI BRACE, ONE LENGTH UNIT INSTALLATION
'Type A' Bracing Unit

(Bracing capacity 0.8KN of braced length of wall).

Fig 3

Top Plate
Studs
Nogging

Brace angle preferably between 40º to 50º, but *no* less than 30º MIN., or more than 60º MAX.

Stud Tie
Stud Tie

1.8m MIN. to 2.7m MAX.
for walls **NOT** more than 3000mm high

Fig 4

Stud tie or SN2 strap nail on each face.

Mini Brace

2/35x3.15mm product nails at each stud & wall plate.

END FIXING

MINI BRACE, TWO LENGTHS UNIT INSTALLATION
'Type A' Bracing Unit
(Bracing capacity 0.8KN of braced length of wall).

Fig 5

Top Plate
Nogging
Bottom Plate

Brace angle preferably 40º to 50º, 30º MIN., 60º MAX.

1.8m MIN. to 2.7m MAX.

Fig 6

2/35x3.15mm product nails MIN. in brace ends.

1/35x3.15mm product nail at each stud crossing.

END FIXING

Flat Strap & Speed Brace

These are tension braces and must be applied in opposing pairs on one side of the frame as in figs 7 & 8. They are *not* housed or slotted in and are attached across the surface of the frame in straight lines at the angles given.

Strap brace must be tensioned by the tensioner provided. Speedbrace is tensioned by hammering flat at stud crossings.

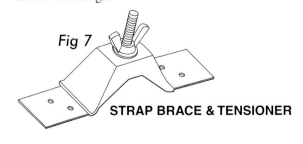

Fig 7

STRAP BRACE & TENSIONER

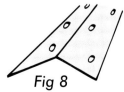

Fig 8

SPEEDBRACE
These can be easier to fit as tensioning is *not* required however they do protrude above the frame a little more than strap.

How to Install Strap & Speed Bracing

Use Strap Brace SB082 for 'Type A' units and SB083 for 'Type B'. Alternatively, use Speedbrace for both 'Type A' and 'Type B' units.

Fixing 'Type B' Bracing

1. Make sure that the wall frame is square.

2. For 'Type B' units, wrap the brace over the plate. Nail the end of the strap brace to the top plate within 150mm of a stud.

3. Lay the strap brace across the frame at the angles illustrated with the unfixed end to wrap around the bottom plate within 150mm.

4. Straighten and partially tighten the strap brace by pulling it down onto the bottom plate.
For 'Type B' units, wrap the brace over the plate.

5. Fix the second length of strap brace or speedbrace in the same manner, diagonally opposing the first.

6. Fit and tighten the tensioners on both frames, with the tensioner facing into the frame. Adjust the tension if necessary to plumb the frame.

7. Nail both braces to every stud crossing.

Using: 2/35x3.15mm Pryda nails for 'Type A' and; 4/35x3.15mm Pryda nails for 'Type B' units.

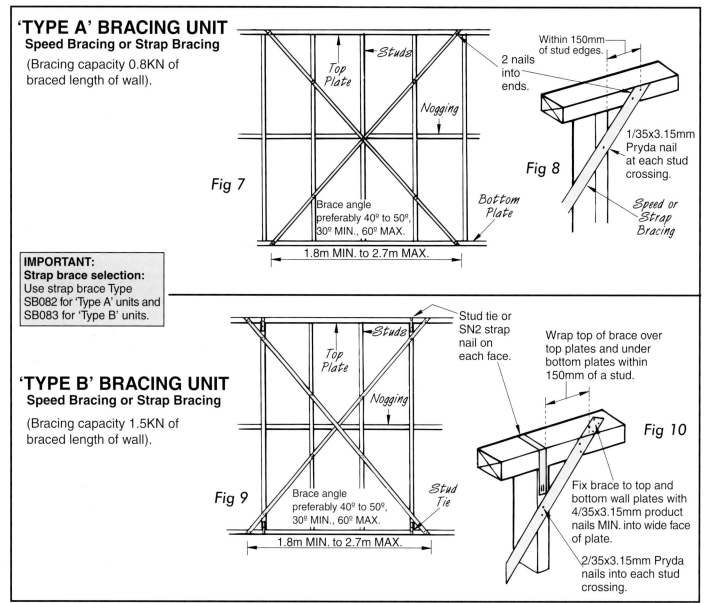

'TYPE A' BRACING UNIT
Speed Bracing or Strap Bracing

(Bracing capacity 0.8KN of braced length of wall).

Fig 7

Top Plate • Studs • Nogging • Bottom Plate

Brace angle preferably 40º to 50º, 30º MIN., 60º MAX.

1.8m MIN. to 2.7m MAX.

Fig 8

Within 150mm of stud edges. • 2 nails into ends. • 1/35x3.15mm Pryda nail at each stud crossing. • Speed or Strap Bracing

IMPORTANT:
Strap brace selection:
Use strap brace Type SB082 for 'Type A' units and SB083 for 'Type B' units.

'TYPE B' BRACING UNIT
Speed Bracing or Strap Bracing

(Bracing capacity 1.5KN of braced length of wall).

Fig 9

Top Plate • Studs • Nogging • Stud Tie

Brace angle preferably 40º to 50º, 30º MIN., 60º MAX.

1.8m MIN. to 2.7m MAX.

Stud tie or SN2 strap nail on each face.

Wrap top of brace over top plates and under bottom plates within 150mm of a stud.

Fig 10

Fix brace to top and bottom wall plates with 4/35x3.15mm product nails MIN. into wide face of plate.

2/35x3.15mm Pryda nails into each stud crossing.

Timber Bracing

See approved plans for specific cross sectional size and stress grading of braces and fastening requirements. These must be free from end splits, or any checks which would adversely affect the fastening of the braces. Pre-drill nail holes at ends. Ensure housings are *not* deeper than specified.

Sheet Bracing

Plywood, hardwood and 0SB sheet bracing is commonly used though the author's preference is plywood. Sheet bracing is often used on external frames on brick veneer as it is concealed inside the cavity. Wall studs must be located to support edges. Where horizontal butt joins occur, a row of noggings should be provided to support both panels. Fastenings must be as required in the approved plans and specifications or in manuf. specifications. Take care that fastenings are 7mm MIN. from panel edges. Locate panels as indicated on approved plans.

Warning: Some imported plywoods may *not* be manufactured according to strict Australian Standards so check their specifications, branding and warranties as delaminating can occur & if not approved may not be accepted by the certifier.

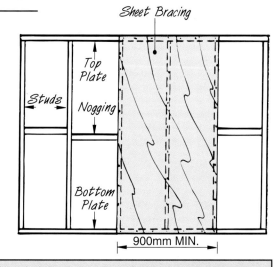

Sheet Bracing

Top Plate • Studs • Nogging • Bottom Plate

900mm MIN.

Note: *Sheet bracing capactiy ranges from 3.4KN/M to over 6.0KN/M. Refer to manufacturers specifications.*

Squaring Frames for Bracing

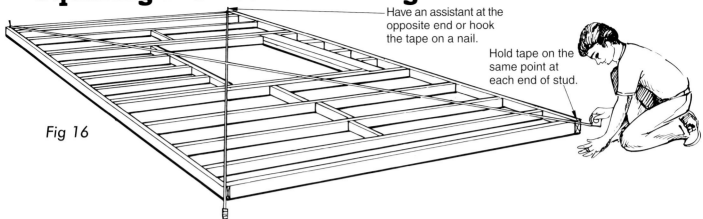

Have an assistant at the opposite end or hook the tape on a nail.

Hold tape on the same point at each end of stud.

Fig 16

Squaring Frame for Bracing

Ensure the bottom plate is straight. It can be lined up to a chalk line or a flooring join. Measure the diagonals of the frame. Hold the tape on the stud marks at the corresponding position in opposite corners (see fig 16). Then rack the frame if necessary until both diagonals correspond. Attach a temporary brace to hold the frame square while the permanent brace is being fitted.

Attaching Permanent Bracing

Attaching Angle Bracing

Angle braces are inserted into a slot to rest flush with the surface.

Step 1 Lay the brace on its proposed position in a straight line and mark a pencil line on the side of the brace at each stud crossing.

Step 2
To insert into a slot:
Adjust the circular saw to the depth of the brace and cut through lines marked. Take care *not* to cut slots deeper than 20mm.

Step 3 Fit brace nailing bottom end fully but only partially drive the nail at the top end. These braces may have to be adjusted when plumbing the frames after erection.

Attaching Speed & Strap Bracing

Speed bracing or flat strap bracing are tension braces only and must be fitted in opposing pairs.

How to Attach

Square the wall frames and attach the braces. Secure the bottom ends but only partially nail the top ends and intermediate stud crossings. Keep the braces in a straight line and at the angle required. All nails are driven fully after the frames have been erected and plumbed. Tension each strap brace after walls are straightened and plumbed. *See manufacturer's literature for further details.*

Attaching Sheet Bracing

The frame is brought into square alignment and a temporary brace attached. The sheet brace can be temporarily attached at this stage or fitted after frame erection, as may be the case when the brace is on the cavity side of brick veneer walls. Ensure the specified fastenings are located at the required spacings.

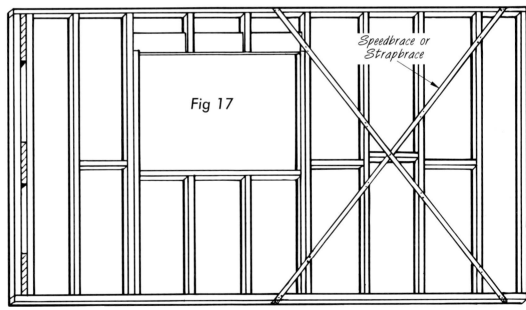

Fig 17

Speedbrace or Strapbrace

Fixing Bracing Walls to the Subfloor

All bracing walls should be fastened to the subfloor according to approved plans. On platform floors, where internal non-load bearing walls contain bracing, an additional joist or noggings may be required underneath for fixing the wall.

Fixing Bracing Walls to the Ceiling or Roof Framing *(see Page 101)*

Lateral Support for External Walls
(see Pages 89&102).

85

Erecting Frames

Erecting Frames

To simplify straightening of bottom plates later on, spring a chalk line on the floor on the inside of the wall plate positions around the perimeter of the house prior to erecting any frames. The four intersecting corners of the sprung chalk line can also be measured diagonally to test for square.

Step 1 If frames are to be erected on a concrete slab, drill any anchor bolt holes through the bottom plates before erecting.

Step 2 Raise the first frame into position. Tack temporary braces to walls to prevent overturning.

Step 3 Continue to erect frames partially nailing them to each preceding frame.

Hint: If when raising frames into position, they have to also be lifted over protruding anchor bolts, place wooden blocks under the bottom plates before lifting the frame to make it easier.

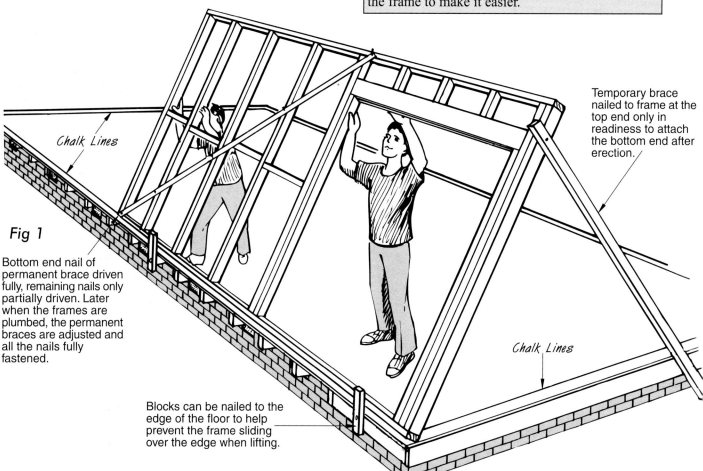

Chalk Lines

Temporary brace nailed to frame at the top end only in readiness to attach the bottom end after erection.

Fig 1

Bottom end nail of permanent brace driven fully, remaining nails only partially driven. Later when the frames are plumbed, the permanent braces are adjusted and all the nails fully fastened.

Blocks can be nailed to the edge of the floor to help prevent the frame sliding over the edge when lifting.

Chalk Lines

Alternative Method for Straightening Bottom Plates

If the chalk line method is not adopted, then the string line and gauge block method can be applied. Fasten the ends of the plates first then attach gauge blocks of equal thickness to all external corners on the edge of the bottom plates. Stretch a string line onto the face of these gauge blocks. Then using a spare guage block, at regular intervals, slide it between the plate and string, adjust the plate, until the block slides neatly between barely touching the string. Nail the bottom plate at these points. Continue this procedure, straigthening the centres and any spring first and working towards the ends.

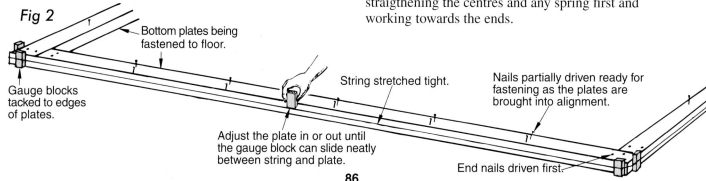

Fig 2

Bottom plates being fastened to floor.

Gauge blocks tacked to edges of plates.

Adjust the plate in or out until the gauge block can slide neatly between string and plate.

String stretched tight.

Nails partially driven ready for fastening as the plates are brought into alignment.

End nails driven first.

Plumbing the Frame

Fig 1

How to Plumb the Frame

Frames can be plumbed by either using a straightedge and level or by a plumb bob and gauge blocks. However, the straightedge and level method is quicker and simpler. The straightedge must be straight and parallel and must reach from the floor to the top plate. The level must be accurate. Either method should be carried out with an assistant who adjusts and secures the temporary braces.

Preparing a Straightedge

Attach small wooden blocks of equal thickness (say 12mm) to each end of the straightedge. When in use, they will rest on the sides of the top and bottom plates. Their purpose is to enable the straightedge to stand free of any spring in the stud (see fig 1).

Step 1 Provide a 90 or 70x35mm temporary brace to every wall, nailing the top ends only to top plates or to a stud under the top plate. Ensure that if a brace extends above the top plate, it will not become an obstruction when constructing the roof. Have a nail partially driven in the bottom of the brace ready for the assistant to attach the moment the level reads plumb as in fig 2.

Step 2 Plumb the internal and external corners. If the wall is out of plumb, release the permanent brace that was temporarily attached when constructing the frame. Then plumb the corner again while the assistant pushes or pulls the frame

into plumb position with the temporary brace. The moment it is plumb, the assistant secures the bottom end of the temporary brace.

Any top plate joins may require adjusting to permit corners to move into plumb position. After the wall is plumbed and temporarily braced, the permanent braces can be fully fastened. Repeat this procedure on all walls.

Straightedge

Ensure level is accurate.

Equal thickness blocks attached to each end to enable the straightedge to stand free of any spring in the studs.

> **JOINING WALLS TOGETHER**
>
> *Internal-to-External & Internal-to-Internal* top plate intersections should be connected as on Pages 73&74.
>
> For straightening and joining the vertical studs drive 2/75x3.75mm MIN. nails @ 800mm centres MIN. down the studs.
>
> When connecting internal bracing walls to external walls follow approved plans and see also Page 76.

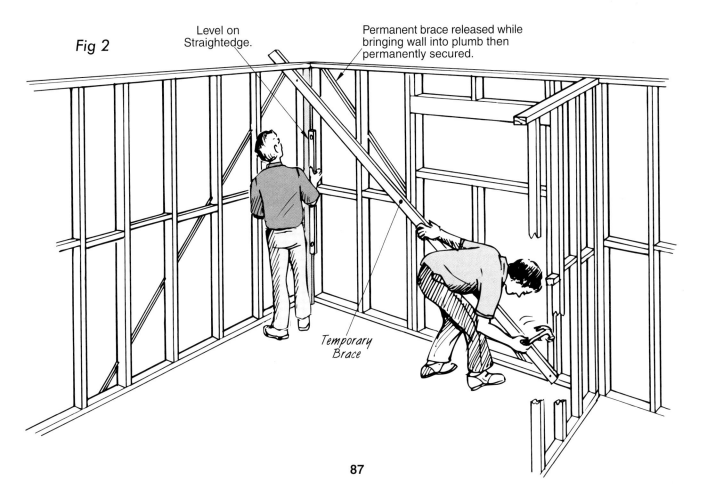

Fig 2

Level on Straightedge.

Permanent brace released while bringing wall into plumb then permanently secured.

Temporary Brace

Straightening Top Plates

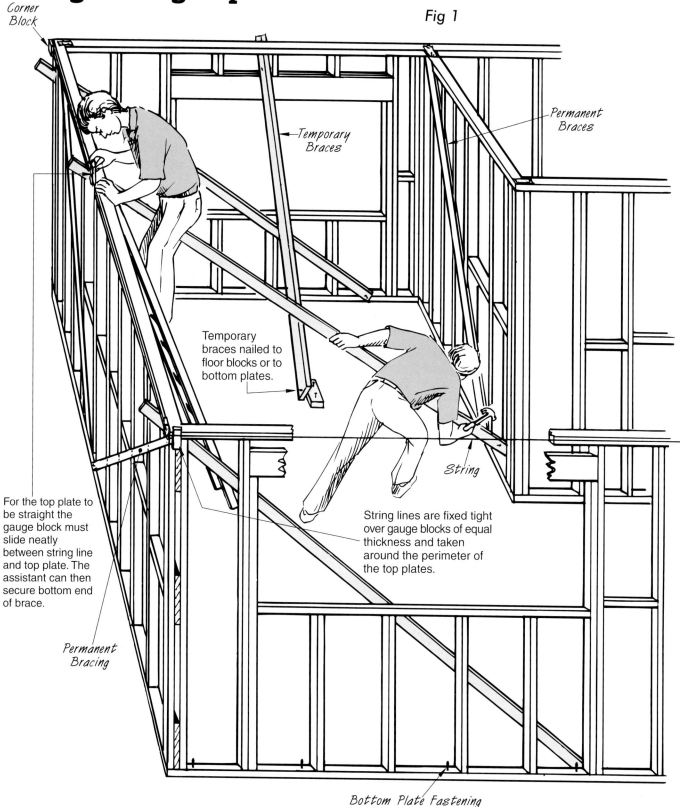

Corner Block

Fig 1

Temporary Braces

Permanent Braces

Temporary braces nailed to floor blocks or to bottom plates.

For the top plate to be straight the gauge block must slide neatly between string line and top plate. The assistant can then secure bottom end of brace.

Permanent Bracing

String

String lines are fixed tight over gauge blocks of equal thickness and taken around the perimeter of the top plates.

Bottom Plate Fastening

After all external and internal corners have been plumbed, top plates are straightened.

Step 1   Fasten all nails or bolts anchoring bottom plates. *(See nominal fastening of bottom plates, Page 86).*

Step 2   Attach 90x35mm or 70x35mm temporary braces to top plates or sides of studs underneath the top plates at points where a spring occurs and/or @1800mm centres approx. Ensure the lower ends

of braces arrive beside the bottom of a stud on an opposite wall bottom plate or to a 90x45mm floor mounted block. *Don't* fix lower ends at this stage.

Step 3   Attach a string line to the outside perimeter of the top plate with equal thickness wood gauge blocks at each corner under the string as in fig 2 page 86.

Using a spare gauge block, slide it between the string line and the side of the top plate while an assistant

Cont.

adjusts the temporary brace in or out until the gauge block slides neatly behind the string. At that moment, secure the bottom end of the temporary brace. Straighten all remaining top plates similarly.

Step 4 All missing noggings can be fitted. Any tie down bolts and straps can be fastened. Retain temporary braces in position until roof is clad.

Lateral Support for External Walls

External walls require lateral support in the form of either adjoining internal partitions at right angles, roof trusses, ceiling joists and ceiling or roof beams. Use of any of these should be at 3m centres MAX.

Alternatively the ceiling frame can be fixed to the external wall frame with ceiling binders at 3m MAX spacings as in either figs 8 & 9 on Page 102.

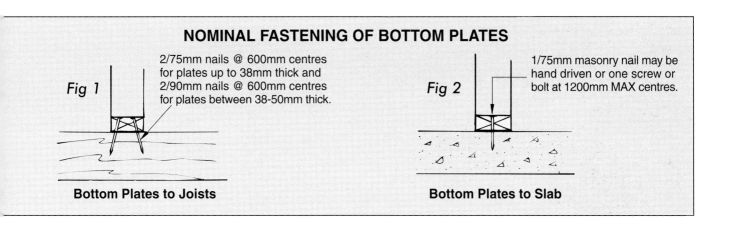

NOMINAL FASTENING OF BOTTOM PLATES

Fig 1
2/75mm nails @ 600mm centres for plates up to 38mm thick and 2/90mm nails @ 600mm centres for plates between 38-50mm thick.

Bottom Plates to Joists

Fig 2
1/75mm masonry nail may be hand driven or one screw or bolt at 1200mm MAX centres.

Bottom Plates to Slab

Some Typical Wall Tie-Down Methods (refer to approved plans for specific methods required)

For tie-down to be effective it must be taken from the roofing down to the footings. This includes roof battens tied to trusses or rafters, trusses or rafters to wall top plates, wall plates to studs, studs to bottom plates, bottom plates to floor framing and floor framing to slab or foundations. House plans will specify the approved method for the specific situation. The following details will familiarise the reader with some common wall to foundation methods.

For some Typical Bearer to Stump Tie-Down Details see Pages 55&56. For Rafter or Truss Tie-Down Details see Page 101.

Tie-Down Methods

Generally the higher the wind category, the stronger the tie-down strength requirement. This is measured in (kN) values. In the overall structure a combination of two or more tie-down methods are frequently specified.

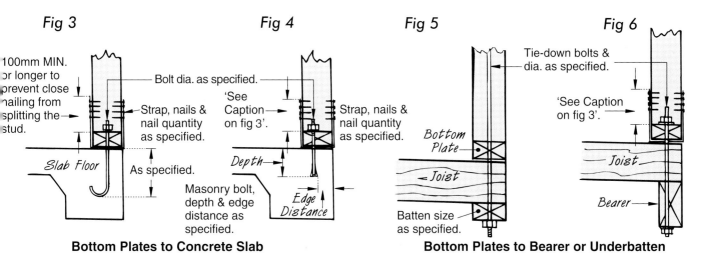

Fig 3

100mm MIN. or longer to prevent close nailing from splitting the stud.

Bolt dia. as specified.

Strap, nails & nail quantity as specified.

Slab Floor

As specified.

Bottom Plates to Concrete Slab

Fig 4

'See Caption on fig 3'.

Strap, nails & nail quantity as specified.

Depth

Masonry bolt, depth & edge distance as specified.

Edge Distance

Fig 5

Bottom Plate

Joist

Batten size as specified.

Fig 6

Tie-down bolts & dia. as specified.

'See Caption on fig 3'.

Joist

Bearer

Bottom Plates to Bearer or Underbatten

STUDS & LINTELS TO PLATES CONNECTIONS

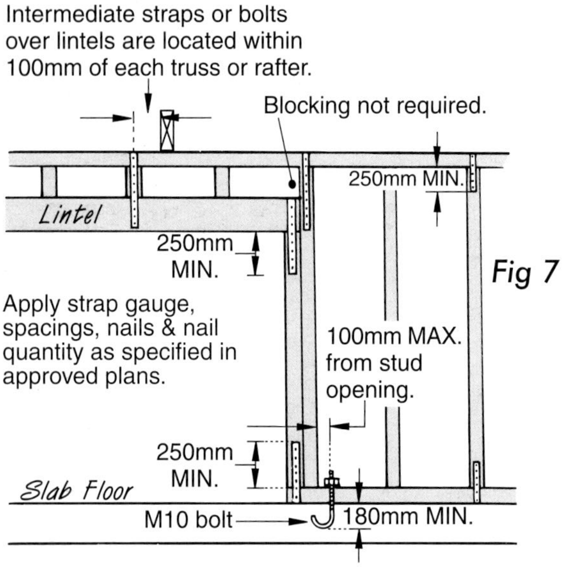

Intermediate straps or bolts over lintels are located within 100mm of each truss or rafter.

Blocking not required.

Lintel

250mm MIN.

250mm MIN.

Fig 7

Apply strap gauge, spacings, nails & nail quantity as specified in approved plans.

100mm MAX. from stud opening.

250mm MIN.

Slab Floor

M10 bolt → 180mm MIN.

Strap Connections

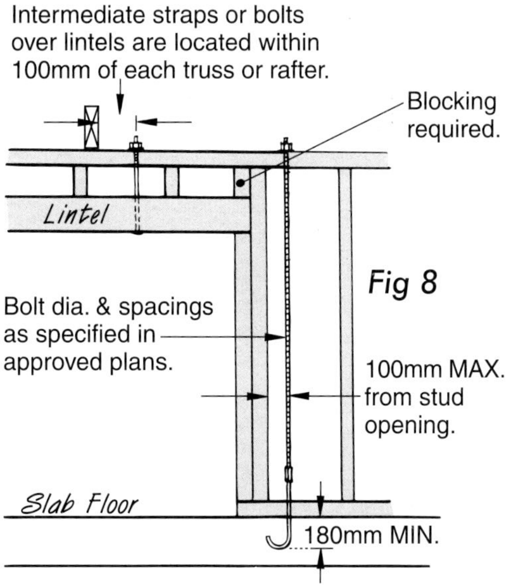

Intermediate straps or bolts over lintels are located within 100mm of each truss or rafter.

Blocking required.

Lintel

Fig 8

Bolt dia. & spacings as specified in approved plans.

100mm MAX. from stud opening.

Slab Floor

180mm MIN.

Bolt Connections

PLY SHEET BRACING AS TIE-DOWN

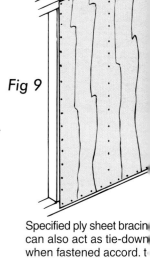

Fig 9

Specified ply sheet bracing can also act as tie-down when fastened accord. t manuf. requirements.

Notching, Trenching & Drilling

Holes drilled in studs and plates *should not* be larger than 25mm dia.

Trenchings or housings *should not* be made adjacent to halved plate joints except at wall junctions.

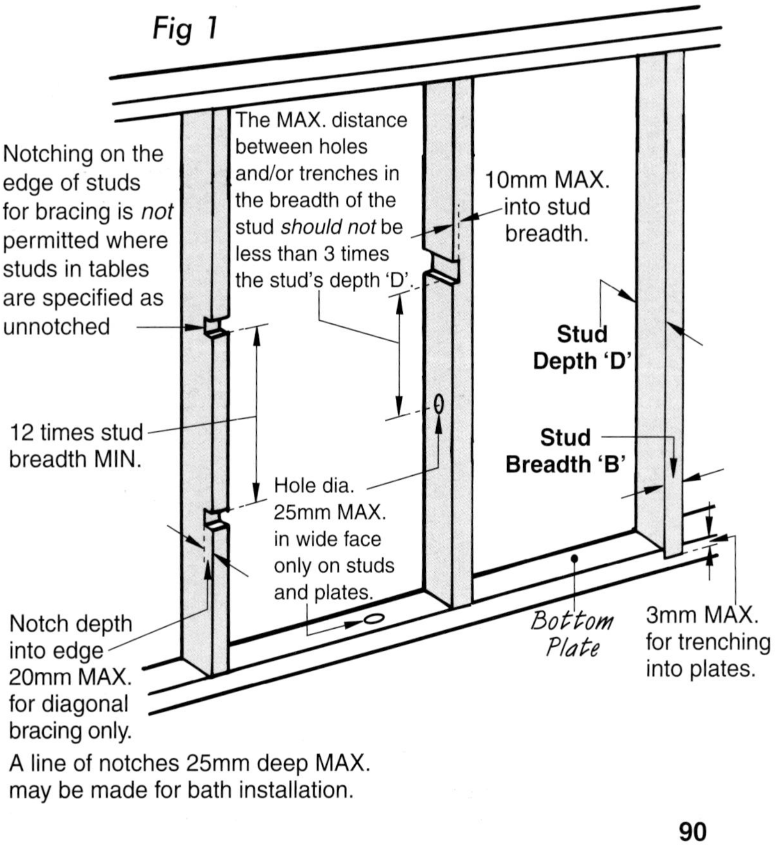

Fig 1

Notching on the edge of studs for bracing is *not* permitted where studs in tables are specified as unnotched

The MAX. distance between holes and/or trenches in the breadth of the stud *should not* be less than 3 times the stud's depth 'D'.

10mm MAX. into stud breadth.

Stud Depth 'D'

Stud Breadth 'B'

12 times stud breadth MIN.

Hole dia. 25mm MAX. in wide face only on studs and plates.

Notch depth into edge 20mm MAX. for diagonal bracing only.

A line of notches 25mm deep MAX. may be made for bath installation.

Bottom Plate

3mm MAX. for trenching into plates.

Straightening Studs

Prior to fixing any external or internal wall linings, any badly sprung studs should be straightened. Wedges are cut approximately 100mm long with a taper from 6mm to nothing.

A saw cut is made on the hollow side of the stud to half the width of the stud maximum at the centre of the spring. The wedges are driven in until the stud is forced straight. The weakened stud is then strengthened by nailing 42x19x600mm long cleats to each side of the saw cut using 4/50mm nails on each side. Do not cut studs beside openings or studs supporting concentrated loads.

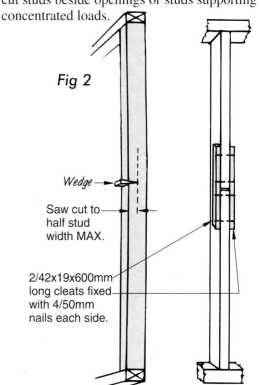

Fig 2

Wedge

Saw cut to half stud width MAX.

2/42x19x600mm long cleats fixed with 4/50mm nails each side.

5

Roof Construction

Prefabricated Roof Trusses

This Chapter deals mainly with prefabricated truss roofs except for Pages 98 & 99 which include cathedral ceilings and near flat roofs.

Prefabricated roof trusses can be used for most roofing designs. In some situations it is more economical and practical to construct a roof partly using trusses and partly conventionally cut on-site.

Important: Manufacturer's instructions must be carefully adhered to with regard to site handling, storage, positioning, anchorage and bracing for their guarantee to remain valid.

Note: For how to cut & pitch conventional cut on-site roofs and to obtain rafter tables & bevels see the companion to this manual *'The Roof Building Manual'.*

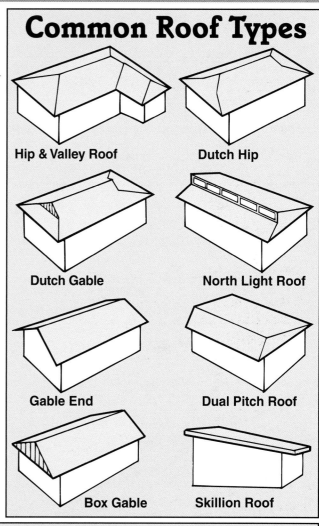

Common Roof Types

Hip & Valley Roof

Dutch Hip

Dutch Gable

North Light Roof

Gable End

Dual Pitch Roof

Box Gable

Skillion Roof

Gable/Eave Finishes

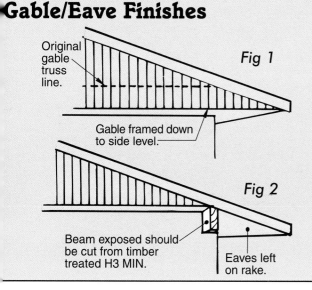

Original gable truss line.

Gable framed down to side level.

Fig 1

Fig 2

Beam exposed should be cut from timber treated H3 MIN.

Eaves left on rake.

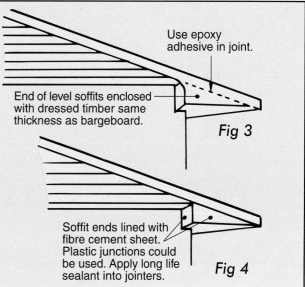

Use epoxy adhesive in joint.

End of level soffits enclosed with dressed timber same thickness as bargeboard.

Fig 3

Soffit ends lined with fibre cement sheet. Plastic junctions could be used. Apply long life sealant into jointers.

Fig 4

Truss Types

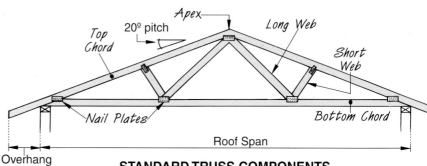

STANDARD TRUSS COMPONENTS

The Hip Truss System

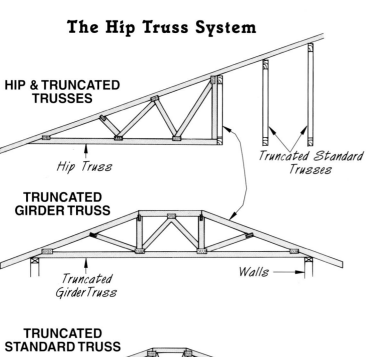

HIP & TRUNCATED TRUSSES

Hip Truss

Truncated Standard Trusses

TRUNCATED GIRDER TRUSS

Truncated Girder Truss

Walls

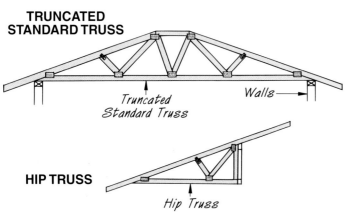

TRUNCATED STANDARD TRUSS

Truncated Standard Truss

Walls

HIP TRUSS

Hip Truss

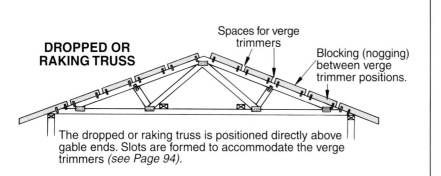

DROPPED OR RAKING TRUSS

Spaces for verge trimmers

Blocking (nogging) between verge trimmer positions.

The dropped or raking truss is positioned directly above gable ends. Slots are formed to accommodate the verge trimmers *(see Page 94).*

Other Truss Designs

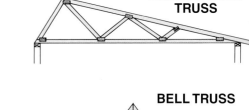

STANDARD TRUSS

Suitable for spans from 9000mm to 12500mm.

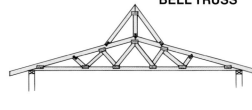

CANTILEVER TRUSS

Cantilever & Overhang Lengths

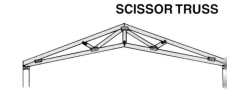

DUAL PITCHED TRUSS

BELL TRUSS

SCISSOR TRUSS

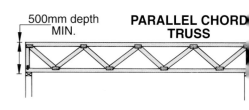

500mm depth MIN.

PARALLEL CHORD TRUSS

BOTTOM CHORD RESTRAINTS

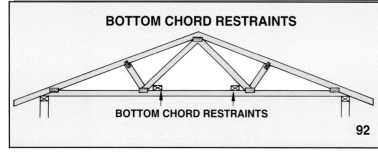

BOTTOM CHORD RESTRAINTS

Bottom chord restraints are required where there is no ceiling, where there are no ceiling battens or where ceiling plasterboard is only nail fastened. Where bottom chords have plasterboard glue and screw fastening, bottom chord restraints are not required. Bottom chord restraints are spaced and fixed according to the manufacturer's specifications. Should this information not be available, provide a 35x70mm @ 3000mm ¢ and fix according to **'Ceiling Binders' Page 102**. Ceiling binders are used to provide lateral support for external walls. Where the approved plans specify ceiling binders, they can also double as bottom chord restraints.

Erecting Roof Trusses

Erecting Standard Trusses on a Gable Ended Roof

On arrival trusses are loaded by the truck crane directly onto the roof. Distribute them into two or three stacks supported by partitions or braced props.

Step 1 Mark off truss positions on wall plates using spacings stipulated by the manufacturer.

Step 2 Erect end trusses 'A' and brace according to fig 1. Plumb and tack in position.

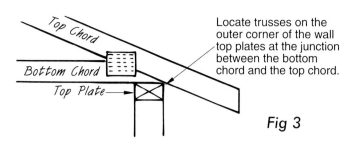

Locate trusses on the outer corner of the wall top plates at the junction between the bottom chord and the top chord.

Top Chord

Bottom Chord

Top Plate

Fig 3

Temporarily brace back to Truss 'A' and tack in place. Continue erecting trusses in this manner.

Step 5 Secure trusses to top plates with triple grips, straps or as specified.

Step 6 After erecting all trusses, attach permanent diagonal braces.

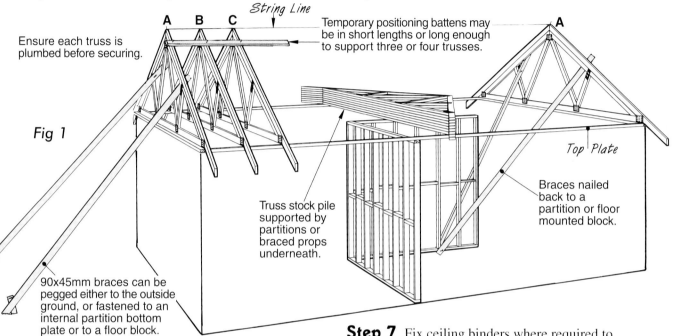

String Line

A B C

Ensure each truss is plumbed before securing.

Temporary positioning battens may be in short lengths or long enough to support three or four trusses.

A

Fig 1

Top Plate

Truss stock pile supported by partitions or braced props underneath.

Braces nailed back to a partition or floor mounted block.

90x45mm braces can be pegged either to the outside ground, or fastened to an internal partition bottom plate or to a floor block.

Braces can be pegged to the ground outside the building. If these braces are too long, they will prove ineffective. In this case, attach them internally.

Step 3 Stretch a string line from the apex of one gable or end truss to the apex of the other (as in fig 1). All intermediate trusses are positioned exactly to the string line.

Step 4 Position and plumb Truss 'B'.

Step 7 Fix ceiling binders where required to bottom cords as in figs 8 or 9, Page 102.

Do not leave truss roofs overnight without anchoring and bracing. Ensure a 10mm gap is maintained between the bottom of the truss and partitions at mid span to allow for deflection. Fix to partitions as on Page 101. Gable ends can finish flush with the end wall or extend past to provide a verge as below.

Step 8 Complete gable end framing *(see Page 96)*.

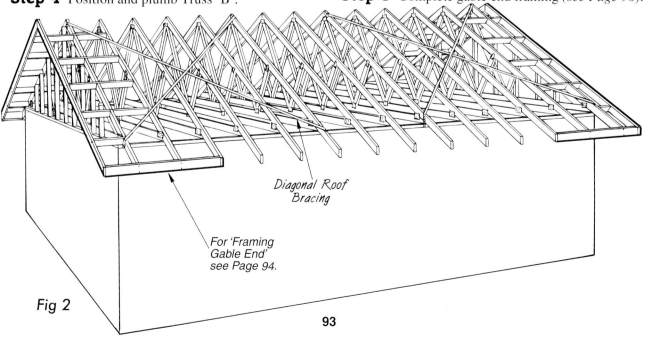

Diagonal Roof Bracing

For 'Framing Gable End' see Page 94.

Fig 2

93

Gable Construction

Dropped Gable Trusses

Dropped trusses contain two top chords. The upper chord is notched to house the verge rafters which cantilever off the lower top cord to support the verge rafters.

Cantilevered Members

Members which are cantilevered to support gables should *not* extend beyond their supports by more than 25% of their permissible span and the backspan should *not* be less than two times the depth of the cantilever.

Step 2 Fix verge rafters in position.

Step 3 Fix cantilevered trimmer and spring a chalk line through the ends of all verge rafters and trimmers and trim their ends.

Step 4 Secure the fly rafter in position.

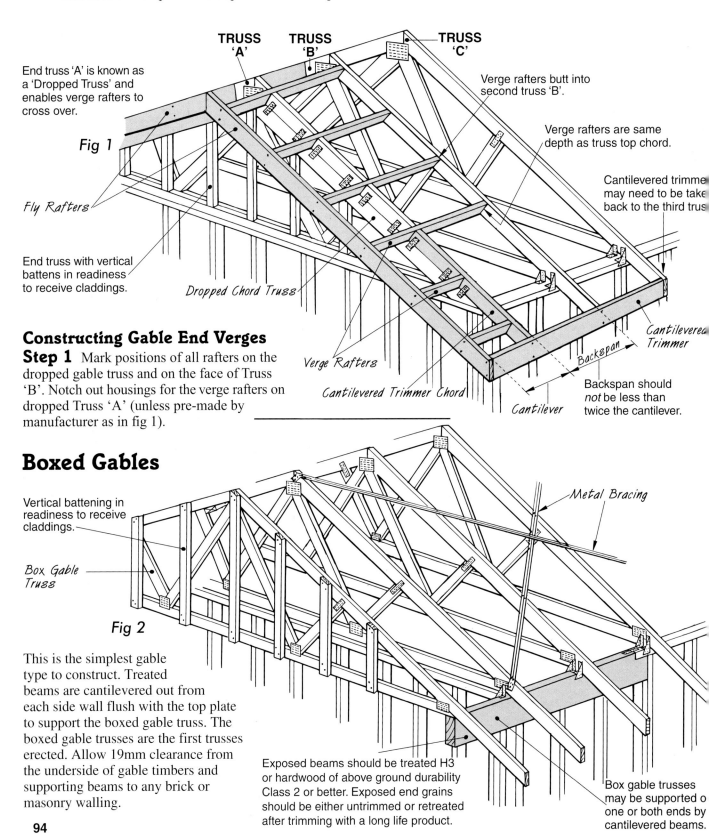

End truss 'A' is known as a 'Dropped Truss' and enables verge rafters to cross over.

Fig 1

Fly Rafters

End truss with vertical battens in readiness to receive claddings.

Dropped Chord Truss

TRUSS 'A' **TRUSS 'B'** **TRUSS 'C'**

Verge rafters butt into second truss 'B'.

Verge rafters are same depth as truss top chord.

Cantilevered trimmer may need to be taken back to the third truss

Verge Rafters

Cantilevered Trimmer Chord

Cantilevered Trimmer

Backspan

Cantilever

Backspan should *not* be less than twice the cantilever.

Constructing Gable End Verges

Step 1 Mark positions of all rafters on the dropped gable truss and on the face of Truss 'B'. Notch out housings for the verge rafters on dropped Truss 'A' (unless pre-made by manufacturer as in fig 1).

Boxed Gables

Vertical battening in readiness to receive claddings.

Box Gable Truss

Fig 2

This is the simplest gable type to construct. Treated beams are cantilevered out from each side wall flush with the top plate to support the boxed gable truss. The boxed gable trusses are the first trusses erected. Allow 19mm clearance from the underside of gable timbers and supporting beams to any brick or masonry walling.

Metal Bracing

Exposed beams should be treated H3 or hardwood of above ground durability Class 2 or better. Exposed end grains should be either untrimmed or retreated after trimming with a long life product.

Box gable trusses may be supported on one or both ends by cantilevered beams.

94

Dutch Gables

Dutch gable design offers an interesting break in an otherwise ordinary hip roof. The hips are terminated short and the ridge extended forward until the desired gable size is arrived at.

The gable end can be constructed using conventional cut on-site rafters while the remaining roof is built using manufactured trusses or trusses can be applied throughout as illustrated. The latter will still require a waling plate for attaching the hip and jack trusses. The end truss supporting the gable will be a dutch hip girder truss and will require noggings for fastening claddings. Cladding could be of flat sheet fibre-cement, horizontal or vertical weatherboards or acrylic render.

Flashing should be carefully applied at the junctions of the hip and jack rafters and the end gable truss.

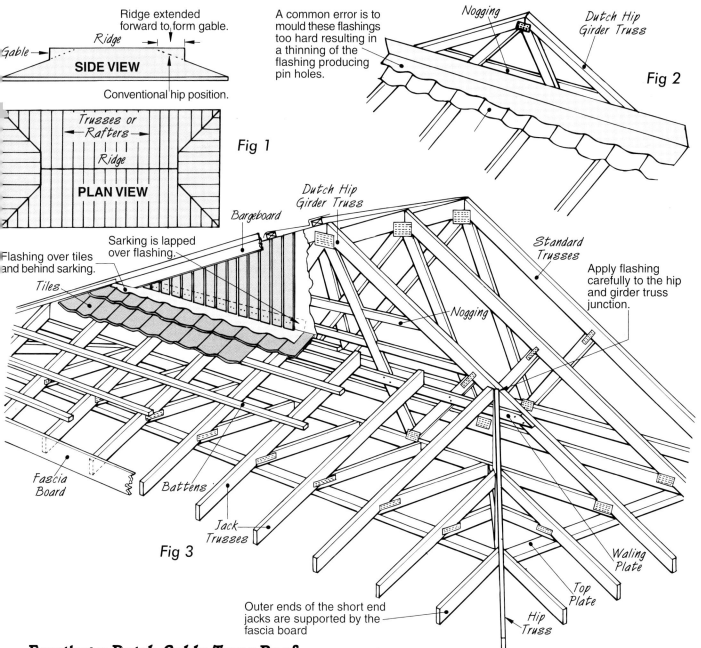

Fig 1

Fig 2

Fig 3

Erecting a Dutch Gable Truss Roof

Step 1 Following the truss spacings specified on the approved plan, mark off the truss positions on the top plate. Ensure accuracy in marking the positions of the dutch hip girder trusses.

Step 2 Erect the dutch hip girder truss and its counterpart at each end where required. Ensure they are plumbed and temporarily braced.

Step 3 Attach the waling plate (if the manufacturer has not already done so).

Step 4 Erect the hip trusses.

Step 5 Erect jack trusses.

Step 6 Install intermediate standard trusses as for gable roofs on Page 93.

Step 7 Ensure all triple grips or anchoring straps are in place and permanent braces attached.

Step 8 Install bottom chord restraints according to manufacturer's requirements.

Hip Roofs

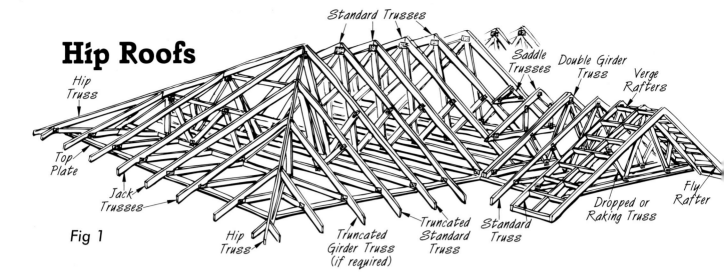

Fig 1

Labels: Hip Truss, Standard Trusses, Saddle Trusses, Double Girder Truss, Verge Rafters, Top Plate, Jack Trusses, Hip Truss, Truncated Girder Truss (if required), Truncated Standard Truss, Standard Truss, Dropped or Raking Truss, Fly Rafter

Erecting Hip End Trusses

Step 1 Mark out truss positions on the wall plates at stations indicated on the manufacturer's plan, or in the approved house plans.

Step 2 Erect truncated standard Truss 'A'. Plumb and temporarily brace.

Step 3 Erect truncated girder Truss 'B'. Plumb and temporarily brace.

Step 4 Erect hip trusses 'C'.

Step 5 Finally erect remaining jack trusses 'E' or jack rafters.

Step 6 Secure all truss ends permanently using specified fixings.

Step 7 After completing the first hip end, operations are then transferred to the opposite end of the roof to erect any other hip ends or gable trusses. A string line is then attached between the two ends at the apex and all intermediate standard trusses 'D' are erected to the string line (see figs 2&3). Trusses and bracing are then fixed as specified.

Important Note: A stud should be positioned directly underneath the truncated girder truss at each end. A large span may require more than one. *See also fig 3, Page 75.*

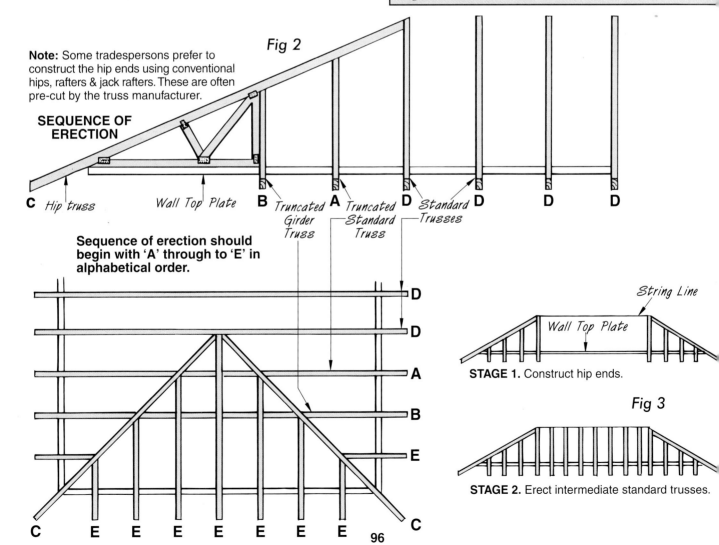

Fig 2

Note: Some tradespersons prefer to construct the hip ends using conventional hips, rafters & jack rafters. These are often pre-cut by the truss manufacturer.

SEQUENCE OF ERECTION

C Hip truss — Wall Top Plate — B Truncated Girder Truss — A Truncated Standard Truss — D Standard Trusses — D — D — D

Sequence of erection should begin with 'A' through to 'E' in alphabetical order.

D
D
A
B
E

C E E E E E E E C

String Line

Wall Top Plate

STAGE 1. Construct hip ends.

Fig 3

STAGE 2. Erect intermediate standard trusses.

Roofs with Valleys

Valley saddle trusses are erected after the installation of the standard trusses to the main roof have been completed.

Erecting Valley Saddle Trusses

Step 1 Erect Truss 'A'. Plumb and temporarily brace.

Step 2 Erect gable Truss 'B'. Plumb and temporarily brace as for a normal gable end as in fig 1, Page 93.

Step 3 Attach a string line on the apex of Truss B through to the main roof as in fig 1. Ensure it is level, square and parallel with the main building. It should also line up with the apex of Truss A. Attach the string to a temporary nogging fitted between the trusses on the main roof.

Step 4 Position remaining trusses fixing battens on the lower side of the saddle trusses for extra support as in fig 1.

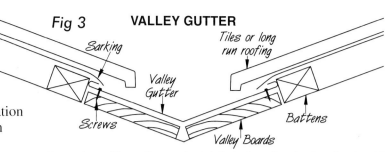

Fig 3 — VALLEY GUTTER

Use a long life gutter material, Colorbond or better. It should be secured with 12x25mm Climaseal Type 17 UPDM seal screws driven either through the valley as high up as possible or driven outside tight up against the valley with the seals tight down onto the valley edges at the sides. Be sure to birdproof any gaps along the valley and at the valley ends.

Step 5 Secure trusses to wall top plates as on Page 101 and attach any specific diagonal wind braces as on Page 100.

Step 6 Complete gable framing if required.

> **Important Notes:**
> **A.** A girder truss at Truss 'A' is used when there is *no* supporting wall below.
> **B.** If the smallest saddle truss is *not* long enough to span the main trusses, install supporting noggings beneath the unsupported ends.

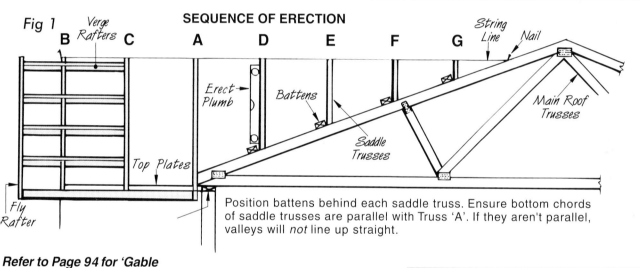

Fig 1 — SEQUENCE OF ERECTION

Position battens behind each saddle truss. Ensure bottom chords of saddle trusses are parallel with Truss 'A'. If they aren't parallel, valleys will *not* line up straight.

Refer to Page 94 for 'Gable End Construction'.

Fig 2

Sequence of erection start with 'A' through to 'G' in alphabetical order.

Ensure saddle trusses are parallel with Truss 'A'

Nogg under end of Truss 'G' if required.

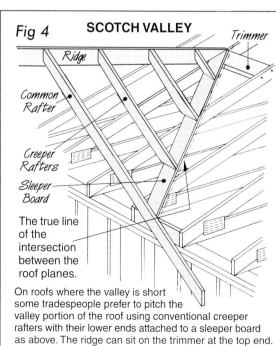

Fig 4 — SCOTCH VALLEY

The true line of the intersection between the roof planes.

On roofs where the valley is short some tradespeople prefer to pitch the valley portion of the roof using conventional creeper rafters with their lower ends attached to a sleeper board as above. The ridge can sit on the trimmer at the top end.

Cathedral Roofs & Ceilings

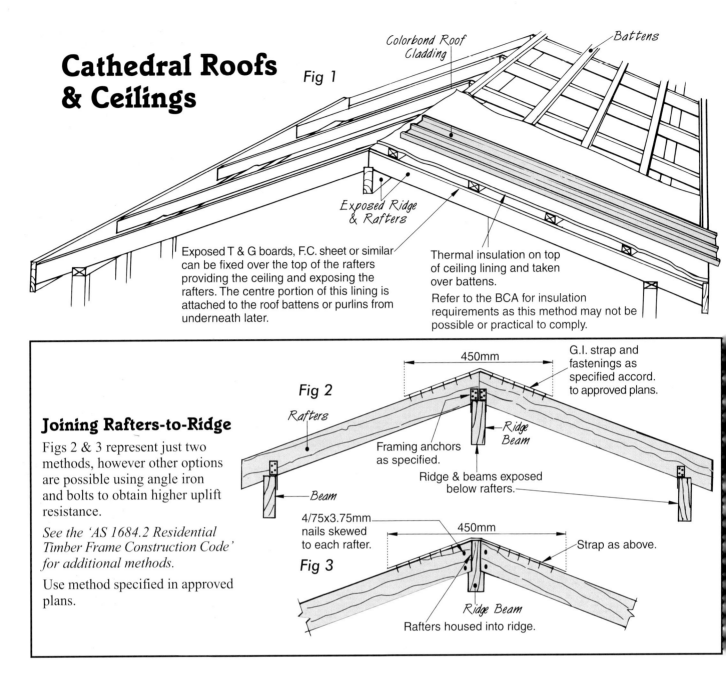

Fig 1

Colorbond Roof Cladding

Battens

Exposed Ridge & Rafters

Exposed T & G boards, F.C. sheet or similar can be fixed over the top of the rafters providing the ceiling and exposing the rafters. The centre portion of this lining is attached to the roof battens or purlins from underneath later.

Thermal insulation on top of ceiling lining and taken over battens.

Refer to the BCA for insulation requirements as this method may not be possible or practical to comply.

Joining Rafters-to-Ridge

Figs 2 & 3 represent just two methods, however other options are possible using angle iron and bolts to obtain higher uplift resistance.

See the 'AS 1684.2 Residential Timber Frame Construction Code' for additional methods.

Use method specified in approved plans.

Fig 2

Rafters

450mm

G.I. strap and fastenings as specified accord. to approved plans.

Framing anchors as specified.

Ridge Beam

Ridge & beams exposed below rafters.

Beam

4/75x3.75mm nails skewed to each rafter.

450mm

Strap as above.

Fig 3

Ridge Beam

Rafters housed into ridge.

Erecting Exposed Rafters

Step 1 Ensure all top plates are in straight alignment. Cut and install the ridge beam.

Step 2 Mark the stations of rafters on the wall top plates and ridge beam. The rafter spacings will be indicated on the house plan. Cut the rafters to length and cut the birdsmouths. It is advisable for the inexperienced to cut and test one pair in position before cutting the remaining rafters.

Step 3 Remaining rafters are cut and attached to the ridge beam and wall top plates. Straps or angle iron and bolts as specified are applied to secure the apex and rafter ends.

Step 4 Ceiling lining is laid on top of rafters and secured.

Step 5 Fix roofing battens according to spacings specified in the approved plans or specifications.

Step 6 Lay insulation across the battens. Seal joins. Attach roof cladding according to approved plans or manufacturer's specifications.

Insulating Sloping Ceilings

This type of roof structure *does not* have an enclosed joisted ceiling space separating the outside air from the interior living space. For this reason, it is essential that insulation be installed between or above the rafters. As most cathedral ceilings have exposed rafters, air-cell type blanket insulation can be laid over and between the battens or purlins as in fig 1.

Important Note: Sheep's wool fibre insulation taken over the battens can cause problems when driving roofing screws as the wool fibres wind around the screws causing bulging of the cladding at the fixing points. Lay wool batts between battens or purlins.

Near Flat Roofs

Constructing Near Flat Roofs

Alternative methods of construction are possible. The one illustrated in figs 1 & 2 is constructed onto walls which are level throughout and the fall in pitch is created either by simply reducing the depth of each rafter or purlin and keeping their bottom sides all on the same level plane to maintain a level ceiling line, as in fig 3 or by the not commonly applied method of varying the rafter or purlin heights. This is achieved by notching the rafters or purlins over the beams or top plates.

> **Note:** Whether a member is a beam or rafter or purlin will depend upon the spans and spacings required of it.

How to Construct either Method

Step 1 Erect the highest and the lowest rafter or purlin.

Step 2 Stretch a string line through their top ends and take the depth measurement remaining at the

Minimum Roof Pitch for Common Claddings
The following roof pitches are recommended for Lysaght Building Product profiles:

Custom Orb Long Run	5° or 1 in 12
Trimdek,	2° or 1 in 30
Spandek,	3° or 1 in 20
Klip-Lok, Hi Ten	
for 0.60mm & 0.48mm thick	1° or 1 in 50
& for 0.42mm	2° or 1 in 30

> **Note:** When constructing roofs at the lowest MIN. pitches, care must be taken that future framing movement or shrinkage *does not* cause ponding. In cylcone areas increase the minimum pitch by one degree.

stations for each remaining rafter or purlin.

Step 3 Notch out the rafters or purlins at their support points or when constructing as in fig 3, purchase rafters or purlins of the correct depth or rip them to fit.

A further method is also common. In fig 4 the walls are constructed on the rake and the rafters are all of the same depth. While the walls are a little more difficult to construct, the roof is very simple.

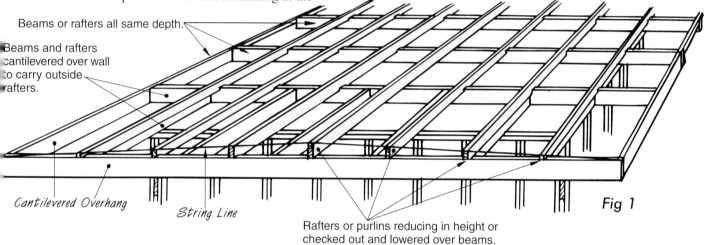

Beams or rafters all same depth.

Beams and rafters cantilevered over wall to carry outside rafters.

Cantilevered Overhang

String Line

Rafters or purlins reducing in height or checked out and lowered over beams.

Fig 1

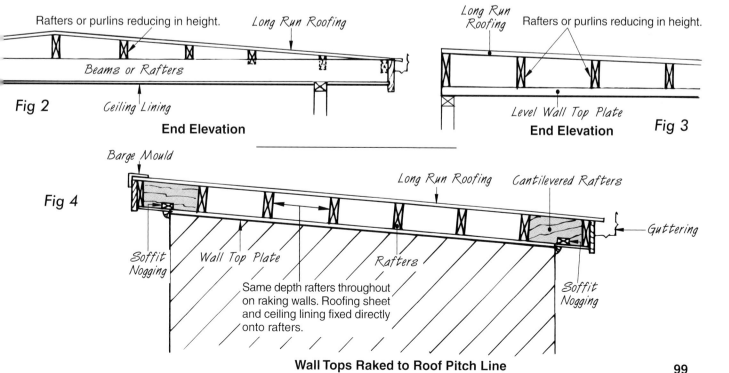

Rafters or purlins reducing in height.

Long Run Roofing

Beams or Rafters

Ceiling Lining

Fig 2

End Elevation

Long Run Roofing

Rafters or purlins reducing in height.

Level Wall Top Plate

End Elevation

Fig 3

Barge Mould

Fig 4

Long Run Roofing

Cantilevered Rafters

Guttering

Soffit Nogging

Wall Top Plate

Rafters

Soffit Nogging

Same depth rafters throughout on raking walls. Roofing sheet and ceiling lining fixed directly onto rafters.

Wall Tops Raked to Roof Pitch Line

Diagonal Roof Bracing

Roof bracing must *not* be neglected and should be applied on the same day the trusses are installed unless temporary braces are applied. Metal or timber bracing can be applied, however timber bracing is not in common usage. Metal bracing is available in flat strap or angled as in speed bracing. These are only effective in tension and therefore must be fitted in opposing pairs (see fig 1). Speed bracing is the easiest to fit. Apply metal bracing in a straight line at approx. 30° to the ridge line and use 30x3.15mm Pryda product nails throughout as these best fit the hole diameter.

> *Note:* Plans should indicate bracing locations and manufacturer's 'Installation Guidelines for Timber Roof Trusses' should also be consulted.

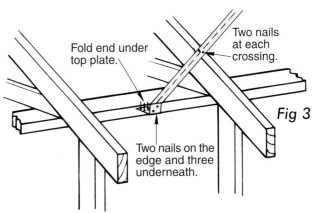

Fold end under top plate.

Two nails at each crossing.

Two nails on the edge and three underneath.

Fig 3

ALTERNATIVE END FIXING SITUATIONS

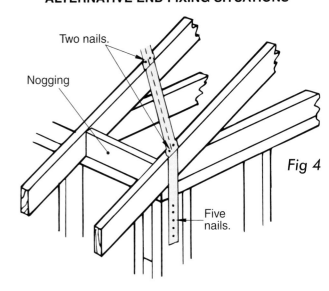

Two nails.

Nogging

Five nails.

Fig 4

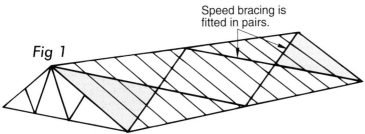

Fig 1

Speed bracing is fitted in pairs.

Roofing battens should run continuous without joins in shaded areas, also brace both sides of roof.

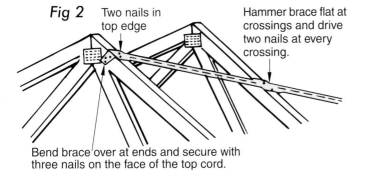

Fig 2

Two nails in top edge

Hammer brace flat at crossings and drive two nails at every crossing.

Bend brace over at ends and secure with three nails on the face of the top cord.

ALTERNATIVE END-TO-END JOINS

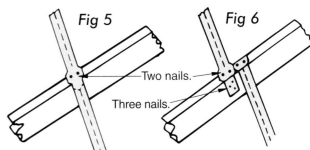

Fig 5

Fig 6

Two nails.

Three nails.

Ends overlapped with three nails through common holes.

Ends side-by-side touching and bent over rafter with three nails into rafter sides.

Roof Truss Fixing

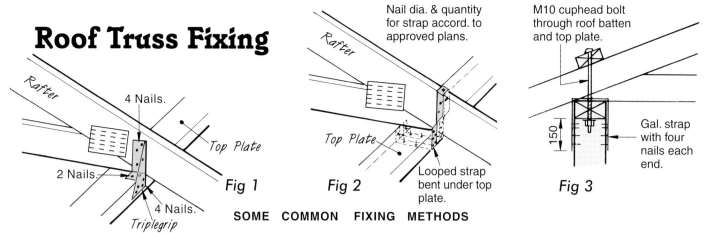

SOME COMMON FIXING METHODS

Fig 1 — Rafter, 4 Nails, 2 Nails, 4 Nails, Triplegrip, Top Plate

Fig 2 — Nail dia. & quantity for strap accord. to approved plans. Rafter, Top Plate, Looped strap bent under top plate.

Fig 3 — M10 cuphead bolt through roof batten and top plate. 150. Gal. strap with four nails each end.

Manufactured roof trusses *should not* be secured by skew nailing alone. Refer to approved plans for specified fixings for the particular situation. If more than one triplegrip is recommended, it may be economical in time and cost to use the strap method in fig 2. This strap has been specifically designed for anchoring truss ends and nail holes are provided. The specifications may require two straps in extreme situations or M10mm bolts are sometimes called for as in fig 3. *Refer to approved plans for specified method.*

Fixing Walls to Ceiling Joists or Truss Bottom Chords

Do not secure partitions to truss bottom chords by driving nails up from the underside of top plates.

Fixing Non-Bracing Walls to Ceiling Joists or Bottom Chords

Non-bracing walls are fixed to truss bottom chords at 1800mm centres with slotted metal top chord brackets as in figs 4 & 5. The slots permit truss deflection.

Allow 10mm MIN. space between bottom chord and top plates.

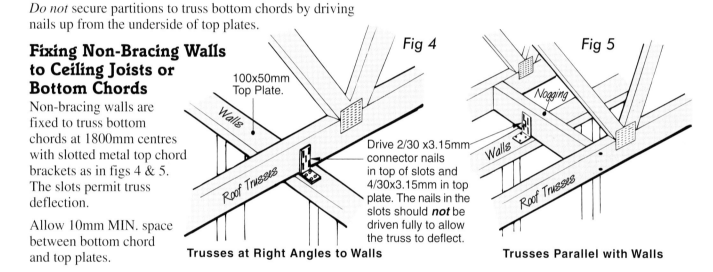

Fig 4 — 100x50mm Top Plate. Walls. Roof Trusses. Drive 2/30 x3.15mm connector nails in top of slots and 4/30x3.15mm in top plate. The nails in the slots should **not** be driven fully to allow the truss to deflect.

Trusses at Right Angles to Walls

Fig 5 — Nogging. Walls. Roof Trusses.

Trusses Parallel with Walls

Attaching Bracing Walls to Ceiling Joists or Truss Bottom Chords

Internal bracing walls should be attached to ceiling joists or truss bottom chords. Examples are shown in figs 6 & 7. Use fastenings according to approved plans. Provide a 10mm MIN. clearance between the underside of joists or bottom chords and wall plates to allow for roof deflection. The truss bottom chords are free to deflect between the blocks.

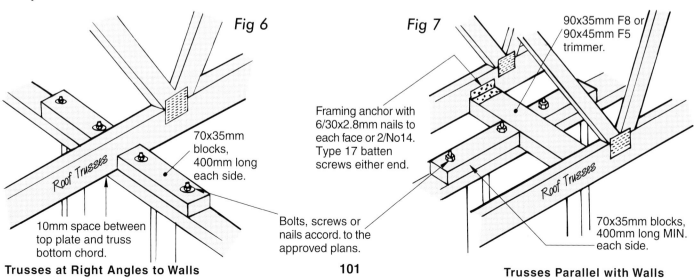

Fig 6 — Roof Trusses. 70x35mm blocks, 400mm long each side. 10mm space between top plate and truss bottom chord. Bolts, screws or nails accord. to the approved plans.

Trusses at Right Angles to Walls

Fig 7 — 90x35mm F8 or 90x45mm F5 trimmer. Framing anchor with 6/30x2.8mm nails to each face or 2/No14. Type 17 batten screws either end. Roof Trusses. 70x35mm blocks, 400mm long MIN. each side.

Trusses Parallel with Walls

Lateral Support for External Walls

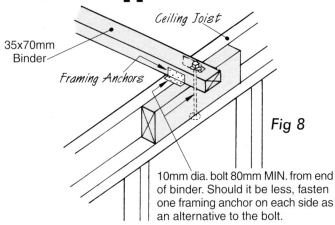

35x70mm Binder

Ceiling Joist

Framing Anchors

Fig 8

10mm dia. bolt 80mm MIN. from end of binder. Should it be less, fasten one framing anchor on each side as an alternative to the bolt.

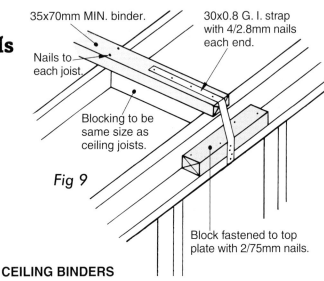

35x70mm MIN. binder.

Nails to each joist.

30x0.8 G. I. strap with 4/2.8mm nails each end.

Blocking to be same size as ceiling joists.

Fig 9

Block fastened to top plate with 2/75mm nails.

CEILING BINDERS

External walls require lateral support in the form of either adjoining internal partitions at right angles, roof trusses, ceiling joists and ceiling or roof beams. Use of any of these should be at 3m MAX centres. Alternatively, the ceiling frame can be fixed to the external wall frame with 35x70mm continuous ceiling binders spaced @ 3m MAX centres apart and fixed to each ceiling joist or bottom chord and to each end wall as in either figs 8 or 9. Where binders are to be used to straighten ceiling joists, use 100x50mm.

Fixing Roof Battens

Various fastening methods are used depending on the design requirement. Figs 10 to 17 are examples of methods commonly used. The tie down strength values of each vary, depending upon the applied fastener, its penetration depth and the strength of the receiving member. Approved plans should specify the required method. Loss of roof including the battens is a common occurrence in storms and high winds. It is essential that the fastening method designated in the approved plans be carefully adhered to. Where fastenings split the timbers, holes should be pre-drilled.

Some Typical Batten Fixing Methods

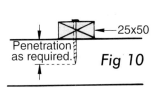

Penetration as required. **Fig 10** 25x50

Nails or screws as specified.

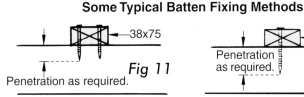

38x75 **Fig 11**

Penetration as required.

Nails or screws as specified.

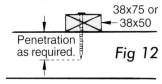

38x75 or 38x50 **Fig 12**

Penetration as required.

Nails or screws as specified.

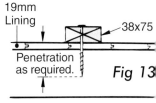

19mm Lining 38x75 **Fig 13**

Penetration as required.

Nails or screws as specified.

Fig 14

1 framing anchor with nails as specified in each leg or, if required 2 framing anchors.

4 Nails **Fig 15**

30x0.8 G.I. strap with nails each end as specified.

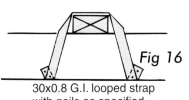

Fig 16

30x0.8 G.I. looped strap with nails as specified.

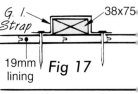

G. I. Strap 38x75

19mm lining **Fig 17**

30x1.8 G.I. strap with nails or screws as specified.

Attaching Fascias & Bargeboards

Fig 18

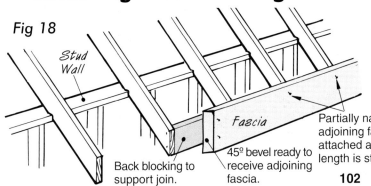

Stud Wall

Fascia

Back blocking to support join.

45º bevel ready to receive adjoining fascia.

Partially nail until adjoining fascias are attached and the entire length is straightened.

BARGEBOARD TO FASCIA JOIN

Fig 19

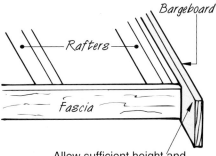

Bargeboard

Rafters

Fascia

Allow sufficient height and overlap to conceal the top outer corner of the guttering.

Prepare Roof for Tiles

For Monier Concrete & Wunderlich Terracotta Tiles

Prior to tiles being laid the roof frame must be prepared to receive the specific cladding. Refer to the following illustrations and captions. You will need to know the batten size to be used by the fixer, better still have a length on-site to use as a guide.

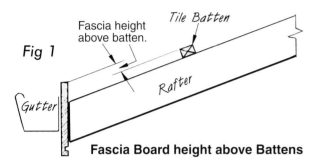

Fascia Board height above Battens

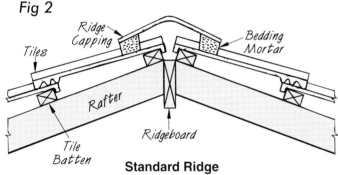

Standard Ridge

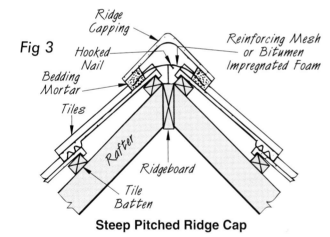

Steep Pitched Ridge Cap

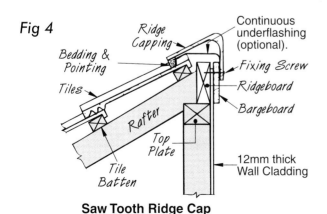

Saw Tooth Ridge Cap

Drawings and details on Pages 105-108; (except fig 2, Page 106) kindly supplied by CSR Roofing

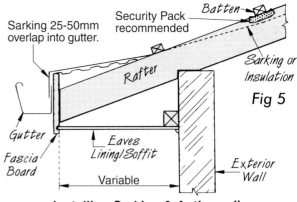

Installing Sarking & Anti-ponding

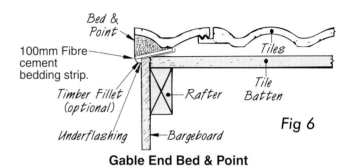

Gable End Bed & Point

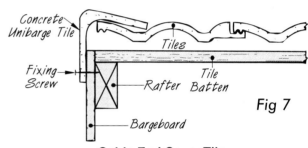

Gable End Cover Tile

Exposed Rafters

In the case of raked ceilings or exposed rafters, ceiling linings, counter battens and sarking are installed on top of the rafters. Counter battens must be fastened over the rafter centre lines to ensure the sarking sag complies with AS/NZS 4200.2. Clearance between the sarking, any insulation material and the celing linings.

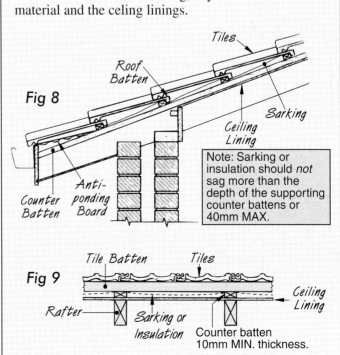

Note: Sarking or insulation should *not* sag more than the depth of the supporting counter battens or 40mm MAX.

Valley Boards

Valley boards must comply with the AS1682.2
Valley boards should *not* extend less than 220mm MIN.
up each slope of the roof. Tiles should overlap each
side of the valley guttering no less than 100mm MIN.
Valley boards should be at least 19mm thick and laid
over the ends of the valley rafters and should be installed
to finish level with the top of the tile batten and must
extend the full width of the valley.

Tapered valley boards 175x25mmx6mm can be used.
The 6mm edge should be placed to the outside of the
valley. Where 38mm thick tile battens are used, a valley
board with an outside thickness of 25mm should be
used.

Where there is a change in direction of a valley, great
care should be taken to ensure that valley boards and
valley irons form a continuous water path to the eaves.
The lip of the valley should, at all points, reach the
height of the roofing battens.

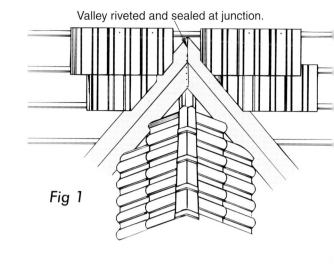

Valley riveted and sealed at junction.

Fig 1

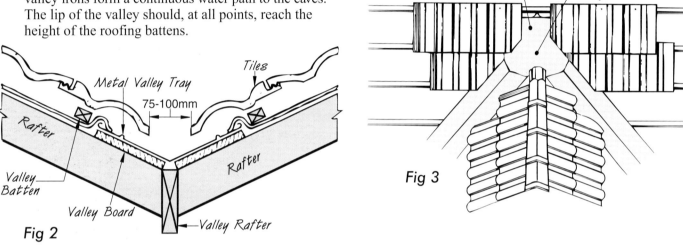

Fig 2

Metal Valley Tray
Tiles
75-100mm
Rafter
Rafter
Valley Batten
Valley Board
Valley Rafter

Valley boards mitred and self supporting at this point.
Saddle Flashing

Fig 3

Dutch Gable & Flashings

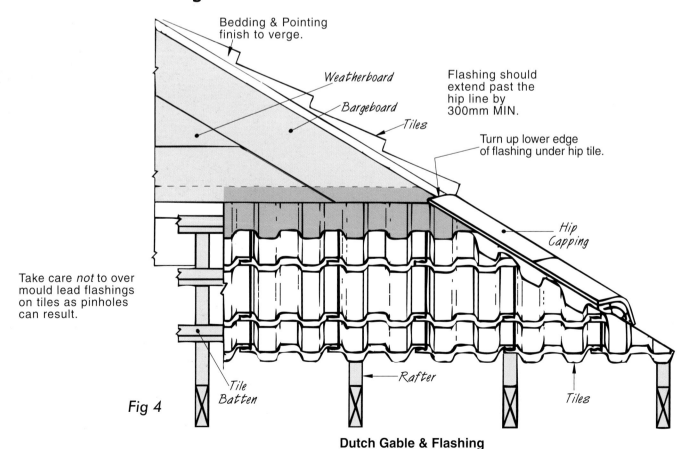

Bedding & Pointing finish to verge.

Weatherboard

Bargeboard

Tiles

Flashing should extend past the hip line by 300mm MIN.

Turn up lower edge of flashing under hip tile.

Hip Capping

Take care *not* to over mould lead flashings on tiles as pinholes can result.

Rafter

Tiles

Fig 4

Tile Batten

Dutch Gable & Flashing
104

Flashings for Tile Roofs

Flashings must be installed to ensure not only that water *does not* enter the interior of the building, but that moisture *does not* come into continual contact with the timber framework eventually causing decay. Where, in future, flashings can not be easily replaced, it is advisable to use colour bonded zinc sheet or copper. Ensure that the metal in flashings and fixings are compatible with the metal onto which they will be in contact to prevent electrolysis. The illustrated examples cover some common situations encountered. Use long life sealant based on silicone or urethane. *Don't use silicone where overpainting in future.*

Flashings for Change of Pitch

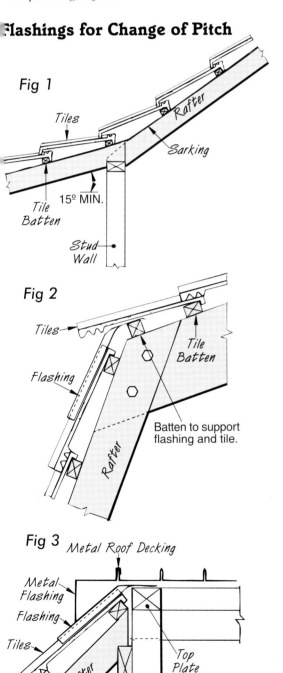

Fig 1

Tiles

Rafter

Sarking

15° MIN.

Tile Batten

Stud Wall

Fig 2

Tiles

Flashing

Tile Batten

Rafter

Batten to support flashing and tile.

Fig 3 Metal Roof Decking

Metal Flashing

Flashing

Tiles

Rafter

Tile Batten

Top Plate

Stud

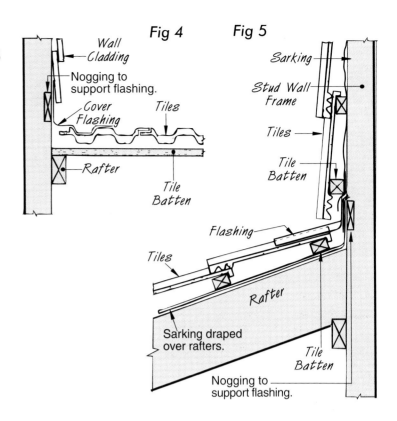

Fig 4

Wall Cladding

Nogging to support flashing.

Cover Flashing

Tiles

Rafter

Tile Batten

Fig 5

Sarking

Stud Wall Frame

Tiles

Tile Batten

Flashing

Tiles

Rafter

Sarking draped over rafters.

Tile Batten

Nogging to support flashing.

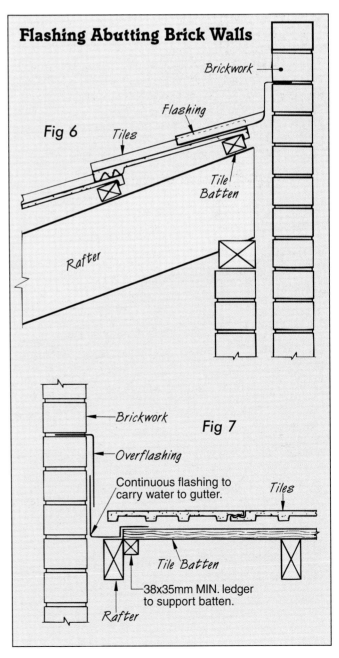

Flashing Abutting Brick Walls

Brickwork

Flashing

Fig 6   Tiles

Tile Batten

Rafter

Brickwork

Overflashing

Fig 7

Continuous flashing to carry water to gutter.

Tiles

Tile Batten

38x35mm MIN. ledger to support batten.

Rafter

Chimney Flashing

Sarking around penetrations in the roof, such as chimneys, shafts, vents and skylight abutments should be trimmed and the edges turned up to divert water around the projections and from under-flashings. The issue of ponding should be considered.

All edges and junctions of finished works should be clean and properly sealed against water penetration.

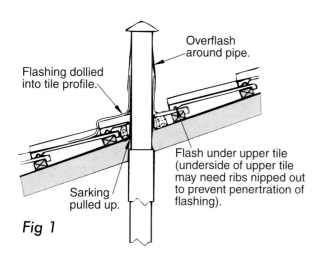

Flashing dollied into tile profile.

Overflash around pipe.

Flash under upper tile (underside of upper tile may need ribs nipped out to prevent penertration of flashing).

Sarking pulled up.

Fig 1

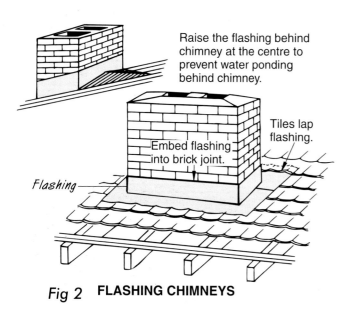

Raise the flashing behind chimney at the centre to prevent water ponding behind chimney.

Tiles lap flashing.

Embed flashing into brick joint.

Flashing

Fig 2 **FLASHING CHIMNEYS**

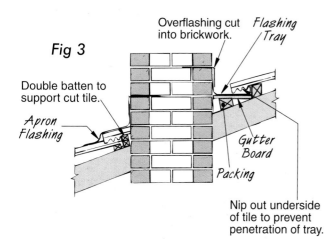

Fig 3

Overflashing cut into brickwork.

Flashing Tray

Double batten to support cut tile.

Apron Flashing

Gutter Board

Packing

Nip out underside of tile to prevent penetration of tray.

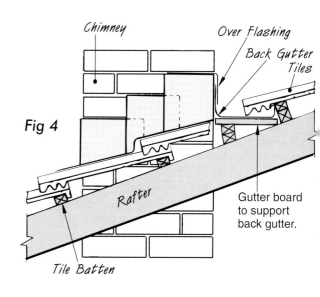

Fig 4

Chimney

Over Flashing

Back Gutter

Tiles

Rafter

Gutter board to support back gutter.

Tile Batten

Use a long life metal such as a heavy gauge copper or stainless steel for the chimney as it can't very easily be replaced.

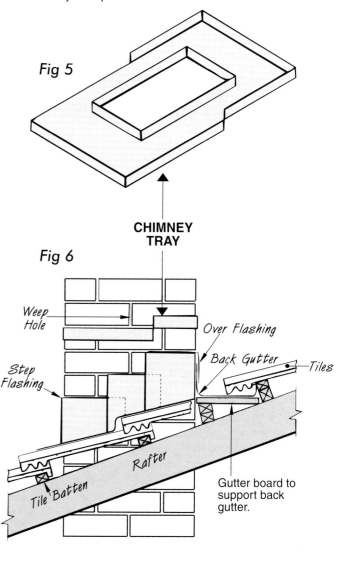

Fig 5

CHIMNEY TRAY

Fig 6

Weep Hole

Step Flashing

Over Flashing

Back Gutter

Tiles

Tile Batten

Rafter

Gutter board to support back gutter.

106

Other Flashings

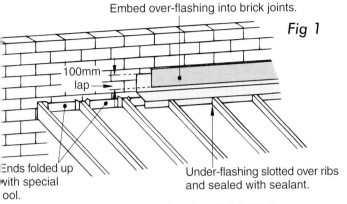

Embed over-flashing into brick joints.

Fig 1

100mm lap

Under-flashing slotted over ribs and sealed with sealant.

Ends folded up with special tool.

Tray Decking Abutting Brick Walls

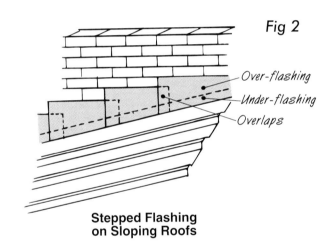

Fig 2

Over-flashing

Under-flashing

Overlaps

Stepped Flashing on Sloping Roofs

Flashings over parapet tops are essential to prevent water penetration.

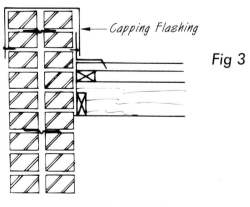

Capping Flashing

Fig 3

PARAPET FLASHING

FLASHINGS TO ROOFS ABUTTING BRICK WALLS

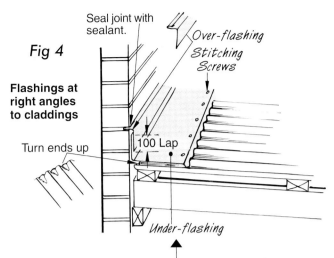

Fig 4

Flashings at right angles to claddings

Seal joint with sealant.

Over-flashing

Stitching Screws

Turn ends up

100 Lap

Under-flashing

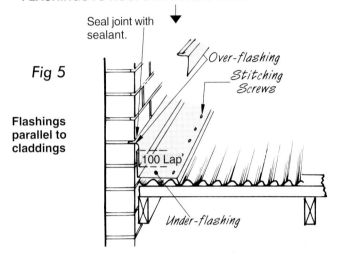

Fig 5

Flashings parallel to claddings

Seal joint with sealant.

Over-flashing

Stitching Screws

100 Lap

Under-flashing

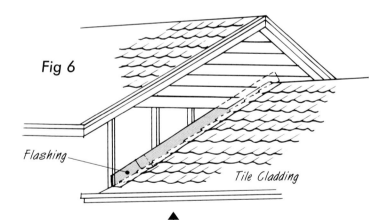

Fig 6

Flashing

Tile Cladding

Flashings on Roofs Abutting Timber Framed Walls

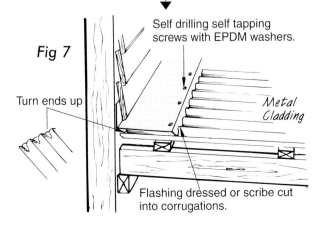

Fig 7

Self drilling self tapping screws with EPDM washers.

Turn ends up

Metal Cladding

Flashing dressed or scribe cut into corrugations.

6

Claddings & Finishing

Weatherboard Sidings

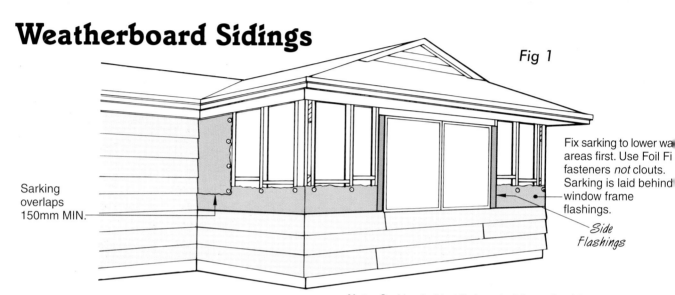

Fig 1

Sarking overlaps 150mm MIN.

Fix sarking to lower wa[ll] areas first. Use Foil Fi[x] fasteners *not* clouts. Sarking is laid behind window frame flashings.

Side Flashings

Weatherboards come in a variety of shapes and materials. These fixing instructions will apply to timber weatherboards, fibre-cement and compressed wood. Additional instruction pamphlets are available from the manufacturers. When ordering, remember that the installed width is less than the overall width because of the lap required (see figs 7 & 8).

Vapour permeable sarking should be attached to walls first, provide 150mm lap at joins and adequate overlap at the joint between the bottom plate and the slab edge. *See* **note** *also*.

Windows and doors are installed, then head and side flashings attached. The ends of weatherboards are cut to butt snugly into the frames to provide a further weather barrier. Some aluminium and timber frames will only permit a butt joint and *not* accept boards inside the frames. *In this case, two methods are possible:* **a).** The boards are cut to fit snugly against the frame and are also embedded in sealant; or **b).** A dressed timber facing of hardwood or treated pine approximately 68x19mm is nailed over the door or window frame edge, fig 5. This overlaps the ends of the weatherboards and provides the necessary weather protection. The end gaps down the edges of the facing can be filled with either a low shrinkage high-build joint sealant or with cut timber scribers as in figs 16-18, Page 110. This latter method is more common for a colonial appearance. Sometimes these gaps are left exposed depending on climate and/or appearance.

Note: Sarking behind timber claddings should permit vapour to pass through *(vapour permeable)* to prevent condensation behind the cladding. On brick veneer walls use waterproof *non permeable* sarking. On weatherboard and brick veneer fasten the top edge at the soffit line on level soffits, *not* to the top plate, to increase air circulation.

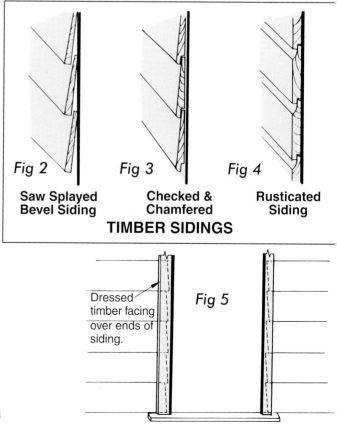

Fig 2
Saw Splayed Bevel Siding

Fig 3
Checked & Chamfered

Fig 4
Rusticated Siding

TIMBER SIDINGS

Dressed timber facing over ends of siding.

Fig 5

Cont.

Fixing Weatherboards

Step 1 Measure the height of the wall to be clad. Take a spacing rod (50x20mm or similar) and transfer the wall height onto the rod. Then divide the rod by the width of the weatherboards being used minus their overlaps (see fig 6). Allow a minimum of 30mm overlap for hardwood, Cypress and treated pine; 25mm for Baltic pine and; 20mm for Western Red Cedar or accord. to manufacturer's spec.

Adjust the boards to equal spacings. Try to adjust the laps so that the lower edge of the first board at lintels above common openings falls in exactly the right place on the head frame or flashings. Where this is not possible, the head weatherboard can be ripped to fit around the opening. *See also Pages 9 & 118.*

Step 2 The lowest board is fixed first. A packing strip is fixed to this board to support it out to the same bevel as the following boards (see fig 9). This packing strip can be attached to the walls or the weatherboard. Spring a chalk line on the studs or sarking through the proposed top edge of the first weatherboard. Ensure it is level and attach the first board to the line. If using timber or compressed wood weatherboards, ensure all end grains receive a coat of wood primer. This first board should lap any base wall or slab edges as in fig 9. Provide 75mm MIN. distance between the lower edges and finished ground level or paving surface.

Step 3 Find the line of the second board by holding the bottom of the spacing rod flush with the bottom of the first board. Transfer the position of the second board from the spacing rod to each end of the wall. Spring a chalk line to these marks. Secure the second board to the chalk line. Repeat this procedure for all remaining boards. When using boards with a rebate, the first board is attached to the chalk line. The following boards are attached with the rebate engaged but not fully home so that there is sufficient space between the boards (on the concealed face) to allow for slight expansion when using timber. However, the edge still needs to be sighted for straight.

Nailing

Use 60x2.8mm gal. plain shank bullet head nails for hardwood or Cypress frames and 60x2.8mm gal. with annular threads for softwood.

Nails should be spaced *not* greater than 650mm horizontally. Drive nails to clear the top edge or tongue of the preceding boards. On Western Red Cedar use only silicon bronze, monel metal, stainless steel or hot dip galvanised nails.

Punch nails below the surface unless otherwise specified. Take care the hammer head is *not* leaving an impression.

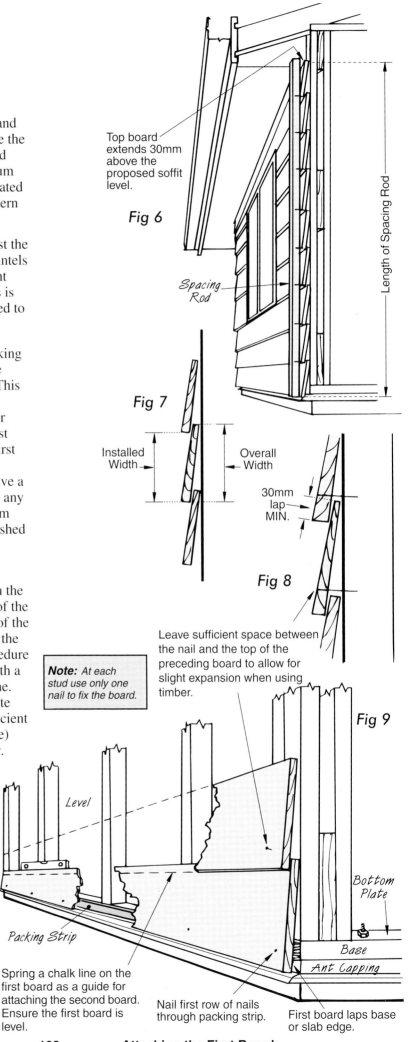

Top board extends 30mm above the proposed soffit level.

Fig 6

Spacing Rod

Length of Spacing Rod

Fig 7

Installed Width

Overall Width

30mm lap MIN.

Fig 8

Leave sufficient space between the nail and the top of the preceding board to allow for slight expansion when using timber.

Note: At each stud use only one nail to fix the board.

Fig 9

Level

Packing Strip

Bottom Plate

Base

Ant Capping

Spring a chalk line on the first board as a guide for attaching the second board. Ensure the first board is level.

Nail first row of nails through packing strip.

First board laps base or slab edge.

Jointing Weatherboards

End-to-End Joins — these should be splay (bevel) cut (see fig 10) and staggered so that joins *don't* arrive in a line above one another.

External Corners — Weatherboards can be mitre and glue jointed or butt joined to a stop-end-batten as in figs 11 & 12.

Internal Corners — Internal corner flashings should be fitted first.

When two walls are clad one row of boards at a time, the end of board 'A', fig 13 is cut square and attached. Board 'B' is firstly positioned resting on two nails. The bevel cut is scribed using either a pair of dividers as is done when scribing mouldings see fig 17 or a small piece of flat material and pencil to follow, transfers the contour of the fixed board onto the loose one (see fig 14). Board 'B' is cut and attached and board 'C' is aligned and notched to fit over board 'B'. The remaining boards are cut in the same manner.

When one wall is boarded up first, all the adjoining boards are scribe cut and notched on their upper edges as in fig 15.

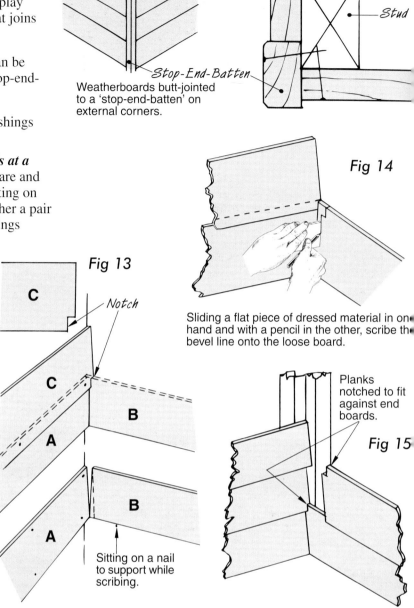

Fig 11

Weatherboards butt-jointed to a 'stop-end-batten' on external corners.
Stop-End-Batten

PLAN VIEW
Fig 12
Stud

Fig 14

Sliding a flat piece of dressed material in one hand and with a pencil in the other, scribe the bevel line onto the loose board.

Fig 13
C
Notch
C
B
A
B
A
Sitting on a nail to support while scribing.

Fig 15
Planks notched to fit against end boards.

Fig 10

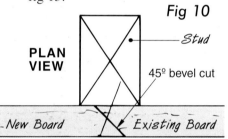

PLAN VIEW
Stud
45º bevel cut
-New Board
Existing Board

End-to-end joins can be butt joined but mitre joining is better.

How to Fit Weatherboard Scribed Mouldings

Use dressed treated pine or hardwood 12 or 19mm thick whichever is necessary.

Step 1 Temporarily secure the moulding over its proposed position allowing the top and bottom ends to extend past their final cut off points.

Step 2 Using a pair of dividers, follow the weatherboards with one leg of the scribers whilst scribing the contour on to the moulding with the other leg (see fig 17). Alternatively, scribe the line using a thin flat piece of material and pencil as in fig 14 (above).

Step 3 Transfer level lines onto the moulding indicating the weatherboard nosings and back edges then join these points together (see fig 18). Scribe any sill or lintel mouldings on to each end.

Step 4 Saw to shape using a fine hand saw or stationary saw. Nail in place tight against the frame by either first pre-drilling through the front edge or in the case of wood joinery, skew nailing through the side.

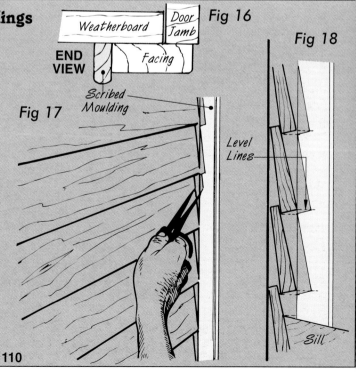

Fig 16
Weatherboard *Door Jamb*
END VIEW *Facing*
Scribed Moulding

Fig 18
Level Lines

Fig 17

Sill

Fixing Scyon™ Linea™ Cladding

For further details including safety instructions refer to James Hardie 'Scyon Linea Cladding' installation manual.

MAXIMUM STUD SPACING

Wind Classification		STUD SPACING	
Non-Cyclonic	Cyclonic	General areas of building (mm)	Within 1200mm of building edges (mm)
N1,N2,N3	C1	600	600
N4,N5	C2,C3	600	450
N6	C4	450	300

Step 1 Ensure framing members are straight and in alignment with one another, install joinery and attach vapour permeable membrane.

Step 2 Install Linea PVC starter strip to the bottom plate as in fig 1. Mark a level line through to indicate the top edge of the first and lowest board. Ensure 20-50mm MIN. overlap of the slab edge as in fig 1. Mark up a storey rod as in fig 6, Page 109 but using Scyon linear spacings and measuring spacings from the level line. Transfer these marks to each end of each wall.

Fastenings — *Concealed Nailing:* 40x2.8mm FC nail Class 3 MIN. flush finished with the surface (see fig 2). *Face Nailing:* (see fig 3) 60x3.15mm bullet head Class 3. Must be driven through both thicknesses of board at laps without predrilling. Stainless steel nails will require predrilling. Use a 3.0mm drill bit. *Gun Nailing:* Must be 40mm MIN. long. Contact supplier for details. *Note:* Keep hand driven nails 20mm MIN. from edges and 50mm for gun nails. *Screw Fastening:* see manuf. literature.

Step 3 Install external and internal corner trims. It is simple to apply Scyon trim or slimline aluminium corners as in figs 4,5,6&7. Fix slimline external corners with 40mm FC nails in indentation provided at 400mm centres to both flanges. When using Scyon trim, apply the flashing first. When using aluminium internal corners, overlap joins 50mm MIN. Upper mould over lower. Fix with 40mm FC nails at 400mm centres on both flanges. Mitre and notch and scribe option is available see manuf. instructions.

End Joins are joined by the tongue and groove provided. A continuous bead of J.H. sealant is applied to the back of the tongue in an upward direction and the boards pushed firmly together.

Step 4 Fix the first board to the studs following the level line. Fix from the centre of boards outwards. *See above for correct fasteners.* Concealed fixing is possible using one nail at the top of boards as in fig 2. However, where a gap occurs between boards, drive a face fixing as in fig 3. Face fixings are punched no more than 2mm. End joins should *not* arrive closer to studs than 100mm. Join as in '*End Joins*' above. All joins are to be staggered.

Around Joinery

If needed install the aluminium window adaptor to the perimeter. Follow figs 8&9. At side reveals insert boards into adaptor and seal with sealant. Seal remaining gaps around joinery with James Hardie Joint Sealant.

Note: The Scyon™ Linea™ cladding must be painted within 90 days.

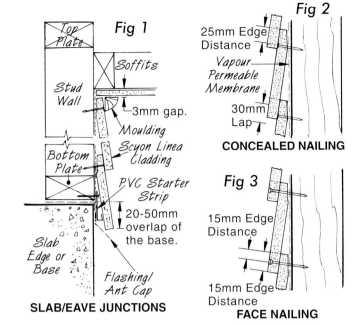

SLAB/EAVE JUNCTIONS

CONCEALED NAILING

FACE NAILING

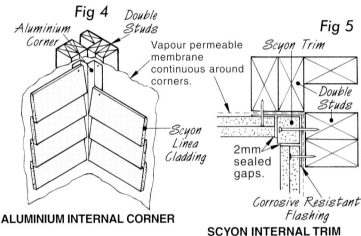

ALUMINIUM INTERNAL CORNER

SCYON INTERNAL TRIM

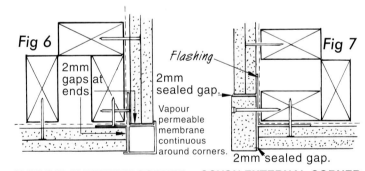

SLIMLINE ALUMINIUM CORNER   **SCYON EXTERNAL CORNER**

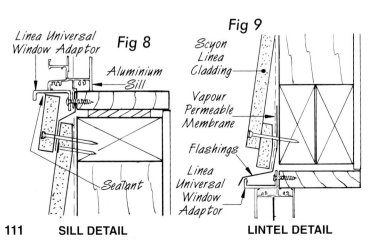

SILL DETAIL   **LINTEL DETAIL**

Fixing Primeline® Cladding

Heritage & Chamfer

Stud Spacing

Must be 600mm MAX. from N1 to N3 wind categories and 450mm MAX. from N4/C1 to N6/C4, but studs within 1200mm of building edge from N5/C3 to N6/C4 must be 300mm MAX.

Step 1 Ensure framing members are straight and in alignment with one another. Install joinery and vapour permeable membrane across the walls accord. to manuf. recommendation.

Step 2 Mark a level line to the perimeter of walls for aligning the top edge of the lowest board. Allow boards to overlap the bearer/base joint as in figs 1&2. Attach internal and external corner moulds (*see figs 3&4, Page 113*).

Step 3 Install the first board into the corner moulds and fasten to each stud and to the level line using 40x2.8mm gal. FC nails. Steel framing 0.55-0.75, use 30mm Buildex Fibre Zip® and 0.8-1.6 framing use 32mm HardiDrive® screws. *For gun nails see James Hardie Installation Manual.* Keep fasteners 40mm MIN. up from the bottom edge (see fig 1). Two fasteners are required in N4/C2 to N6/C4 localities (see fig 2).

End Joins are staggered and can be made on or off the stud.

Joins on the Stud are better and require an offcut of stud material to be attached to the side of the stud. This will necessitate internal linings be carried out after cladding is fixed. The join is made between the stud and offcut. Drive fastenings 20mm each side of the joint after pre-drilling hole (see fig 3).

Joins off the stud Attach an Off-Stud-Clip to the top edge of the fixed board to support following boards as in fig 4 then fasten boards to studs. *Not recommended in coastal areas.*

Step 4 Continue to install boards. Ensure before fastening that all boards fit tightly onto those below.

At Joinery boards are inserted into the frames and preferably bedded into long life sealant (see fig 6). *For alternative treatment, see manuf. instructions.*

At Lintels a flashing should be provided where there is insufficient eaves protection (*see fig 7 below and fig 3, Page 118*).

At Soffits the last board may require ripping to width (*see fig 7*).

Step 5 Fill gaps between joinery frames using James Hardie Gap Sealant and tool smooth.

Note: Primeline cladding must be painted within 90 days.

For further details including safety instructions refer to James Hardie 'External Cladding Installation' instructions.

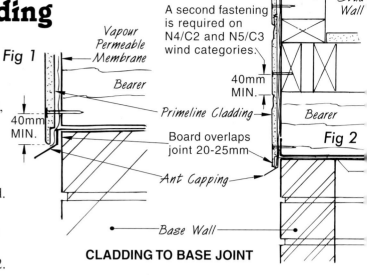

CLADDING TO BASE JOINT

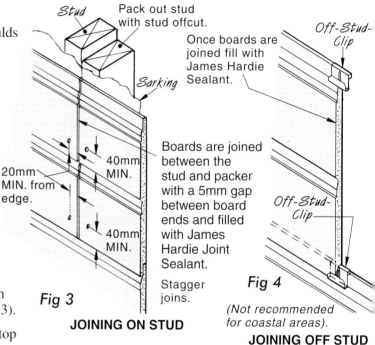

Fig 3 — **JOINING ON STUD**

Fig 4 — (*Not recommended for coastal areas.*) **JOINING OFF STUD**

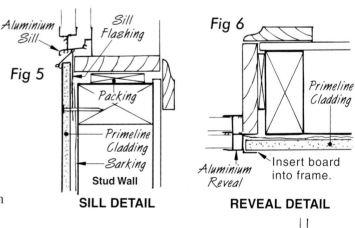

Fig 5 — **SILL DETAIL**

Fig 6 — **REVEAL DETAIL**

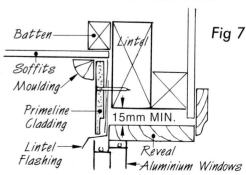

Fig 7 — **LINTEL DETAIL**

Summit & Newport

Stud Spacing

Refer to stud spacing for 'Heritage & Chamfer', Page 112.

Step 1 Ensure framing members are straight and in alignment with one another. Install joinery and vapour permeable membrane across the walls accord. to manuf. recommendation.

Step 2 Mark a level line to the perimeter of the walls for aligning the top edge of the lowest board. Allow boards to overlap the bearer/base joint as in figs 1&2. Attach PVC starter strip as in fig 1 and internal and external corner moulds (see figs 3 & 4).

Step 3 Install the first board into the corner moulds and fasten to each stud and to the level line using fasteners selected from 'Step 3 for Heritage & Chamfer', Page 114. Keep fasteners 20mm MIN. down from the top edge on Summit and 15mm MIN. for Newport as in figs 5 & 6.

End Joins are staggered and can be made on or off the stud. On-the-stud joins are better and will require blocking behind. This will necessitate internal linings be carried out after cladding is fixed. Allow a 5mm gap between boards. This is sealant filled later.

Joins on the Stud require an offcut of stud material to be attached to the side of the stud. The join is made between the stud and offcut. Slide the PVC spline to extend 100mm MIN. into the adjoining board. Drive fastenings 20mm each side of the joint into predrilled holes (see fig 7).

Joins off the stud The join can arrive anywhere between studs but 100mm MIN. away from a stud to receive the spline (see fig 8). Slide the spline 100mm onto the adjoining board and fasten boards to studs as above.

Step 4 Continue to install boards. Insert the PVC spline and ensure before fastening that all boards fit tightly onto those below.

At Joinery (*Similar to 'Heritage & Chamfer', see figs 5,6&7, Page 112*) boards are inserted into the frames and preferably bedded into long life sealant. *For alternative treatment see manuf. literature.*

At Lintels a flashing should be provided where there is insufficient eaves protection (*see fig 7, Page 112 and fig 3, Page 118*).

At Soffits the last board may require ripping to width (*see fig 7, Page 112*). A packing strip will most likely be necessary for Summit & Newport.

Step 5 Fill gaps between joinery frames and tool smooth.

> **Note:** Primeline cladding must be painted within 90 days.

> *For further details including safety instructions refer to James Hardie 'External Cladding Installation' instructions.*

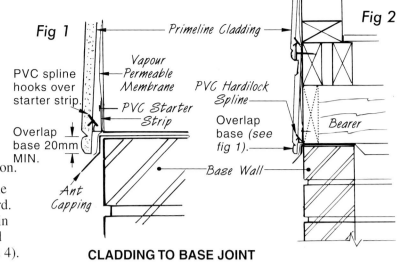

Fig 1 — Primeline Cladding · Vapour Permeable Membrane · PVC spline hooks over starter strip · PVC Starter Strip · Overlap base 20mm MIN. · Ant Capping

Fig 2 — PVC Hardilock Spline · Overlap base (see fig 1). · Bearer · Base Wall

CLADDING TO BASE JOINT

Fig 3 — Vapour Permeable Membrane · Flashing · Don't snap outer piece fully home until full height of weatherboards are fixed in position.

INTERNAL CORNER MOULD

Fig 4 — Flashing · Before inserting outer piece, use pliers to break off teeth along score lines.

EXTERNAL CORNER MOULD

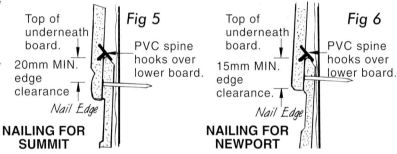

Fig 5 — Top of underneath board. · PVC spine hooks over lower board. · 20mm MIN. edge clearance · Nail Edge

NAILING FOR SUMMIT

Fig 6 — Top of underneath board. · PVC spine hooks over lower board. · 15mm MIN. edge clearance. · Nail Edge

NAILING FOR NEWPORT

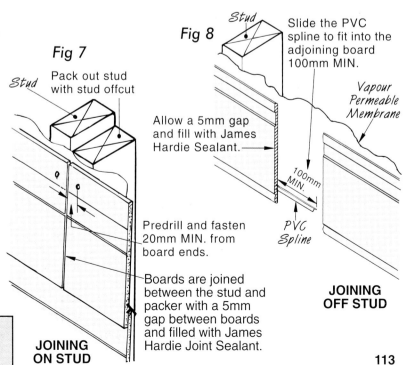

Fig 7 — Stud · Pack out stud with stud offcut · Predrill and fasten 20mm MIN. from board ends. · Boards are joined between the stud and packer with a 5mm gap between boards and filled with James Hardie Joint Sealant.

JOINING ON STUD

Fig 8 — Stud · Slide the PVC spline to fit into the adjoining board 100mm MIN. · Vapour Permeable Membrane · Allow a 5mm gap and fill with James Hardie Sealant. · 100mm MIN. · PVC Spline

JOINING OFF STUD

Installing Villaboard® Lining to Walls

For further details including safety instructions see the James Hardie Villaboard Installation instructions.

Villaboard can be flush jointed similar to plasterboard and is a moisture resistant sheet. As such is the preferred lining for use in wet areas and as a ceramic tile substrate. Steel or timber studs must *not* exceed 600mm spacing. Where sheet joins occur, the studs or noggings must be 38mm MIN. thick. Additional noggings are required above wall/floor flashing or junctions or shower bases for attaching the villaboard. This avoids the need to make nail holes in the waterproof area below. Shower trays or bases are fitted before lining (see Pages 138&139). Use gal. FC nails or screws.

Step 1 Ensure framing members are straight and true.

Step 2 While sheets can be fixed vertically. Horizontal installation is preferred. Joins must occur on a stud or row of noggings. Position the first sheet level and 6mm off the floor; use packers. This gap allows for frame shrinkage.

Fastenings — *To Timber:* use 30x2.8mm gal. FC nails with head dia of 6mm MIN. *For gun nailing, see Villaboard Installation Manual.* *To Steel:* Fix 6mm sheet with 20mm Buildex Fibre Zips® screws for .55-.75 BMT frames (use 30mm Fibre Zips® screws for 9 and 12mm sheets). For .75-1.6 BMT frames use 22mm HardiDrive® screws and 32mm screws for 9 and 12mm sheets.

Keep fastenings 12mm from edges and 50mm away from corners.

Step 3 Fix sheet joins in a staggered pattern (see fig 1).

Untiled Walls (to be painted) fix with fasteners (see fig 2) or a combination of fasteners and construction adhesive (see fig 3) .

Tiled Walls fix with fasteners accord. to fig 4. *For 'Fixing to Masonry' and maximum tile thickness see manuf. literature.*

Jointing Joints are flush finished using James Hardie Base and Top Coat compounds. For all jointing use only paper tape *not* woven linen or self adhesive fibreglass.

Vertical Control Joints
In Timber Frames: *General Use* — 7.2m spacing. *Tiled Walls:* 4.2m.
In Steel Frames: *General Use* — .55-.75 BMT steel 9.0m and 0.8-1.6 BMT 6.0m. *Tiled Walls:* 4.8m for all thickness steel. Horizontal control joints in walls are required at 3.6mm MAX. centres. When sheeting vertically, a horizontal control joint is required at the sheet end when using sheets shorter than 3.6m in length.

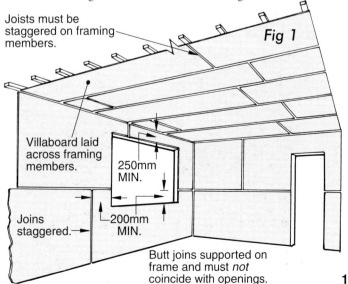

Joists must be staggered on framing members.

Fig 1

Villaboard laid across framing members.

250mm MIN.

200mm MIN.

Joins staggered.

Butt joins supported on frame and must *not* coincide with openings.

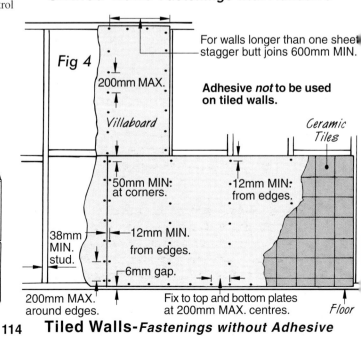

Fig 2

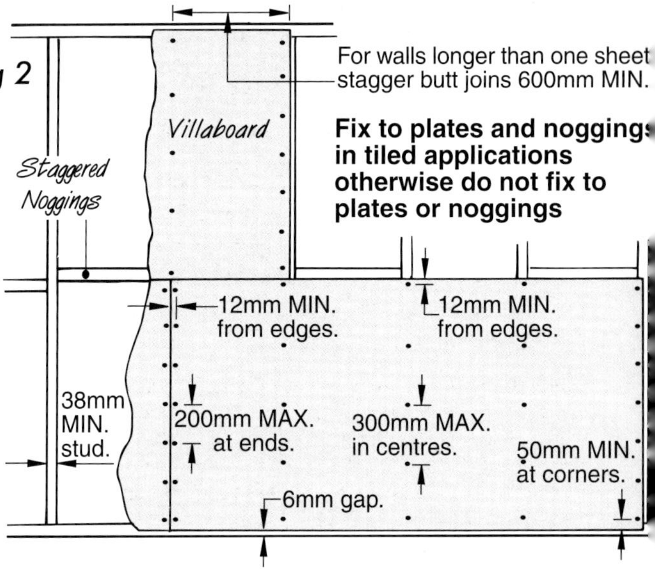

Staggered Noggings

Villaboard

For walls longer than one sheet stagger butt joins 600mm MIN.

Fix to plates and noggings in tiled applications otherwise do not fix to plates or noggings

12mm MIN. from edges.

12mm MIN. from edges.

38mm MIN. stud.

200mm MAX. at ends.

300mm MAX. in centres.

50mm MIN. at corners.

6mm gap.

Untiled Walls-*Fastenings without Adhesive*

Fig 3

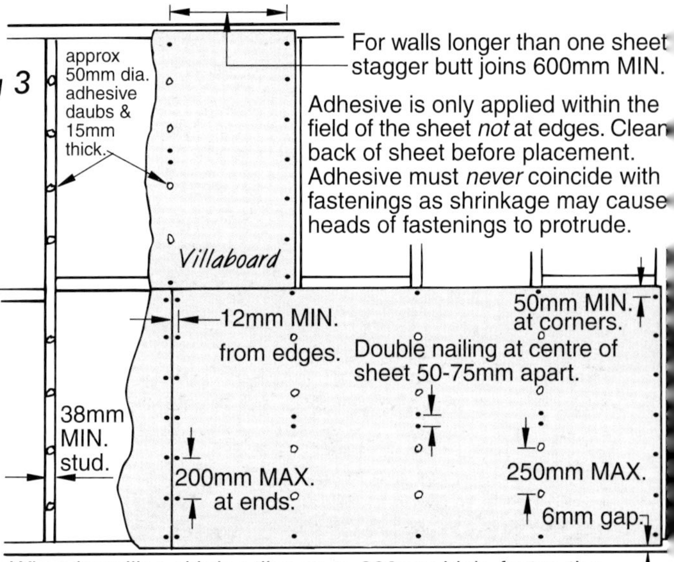

approx 50mm dia. adhesive daubs & 15mm thick.

Villaboard

For walls longer than one sheet stagger butt joins 600mm MIN.

Adhesive is only applied within the field of the sheet *not* at edges. Clean back of sheet before placement. Adhesive must *never* coincide with fastenings as shrinkage may cause heads of fastenings to protrude.

12mm MIN. from edges.

50mm MIN. at corners. Double nailing at centre of sheet 50-75mm apart.

38mm MIN. stud.

200mm MAX. at ends.

250mm MAX.

6mm gap.

When installing skirting tiles up to 300mm high, fasten the bottom edge to the bottom plate at 200mm MAX. centres. All other fasteners are *not* used on plates or noggs.

Untiled Walls-*Fastenings with Adhesive*

Fig 4

200mm MAX.

Villaboard

For walls longer than one sheet stagger butt joins 600mm MIN.

Adhesive *not* to be used on tiled walls.

Ceramic Tiles

50mm MIN. at corners.

12mm MIN. from edges.

38mm MIN. stud.

12mm MIN. from edges.

6mm gap.

200mm MAX. around edges.

Fix to top and bottom plates at 200mm MAX. centres.

Floor

Tiled Walls-*Fastenings without Adhesive*

Installing Harditex® Base Sheets (Blue Board)

Maximum Stud Spacing for 7.5mm sheets:
Space studs at 600mm centres in N1-N3 wind localities 450mm centres in C1/C2/N4 and for N5/C3-N6/C4 300 centres within 1200mm of external corners elsewhere 450 centres.

Step 1 Ensure framing members are straight and in alignment. Use a 42mm width timber stud at sheet joins. Use 900mm wide sheets for 450mm stud centres and 1200mm wide sheets for 600mm stud centres. Install joinery and vapour permeable sarking.

Fastenings — ***To Timber:*** use 30x2.8mm gal. FC nails. *For gun nailing, see Harditex Installation Manual.*
Space fasteners at 200mm centres for N1/N3/C1 and in N4/C2/N5/C3 wind localities 150mm centres within 1200mm of external corners and 200mm elsewhere and in N6/C4 localities 125mm apart within 1200mm of external corners and 150mm elsewhere. *Note:* Adhesives must *not* be used.

Step 2 Sheets must be fixed vertically commencing from a corner and butted hard together, except where movement joints are required. The bottom edge must *not* come into frequent contact with moisture (see fig 2). Ensure 150mm MIN. ground clearance and 50mm MIN. between decking, concrete or paving.

Step 3 Locate fasteners at centres given in fig 1, when *not* used as bracing. Where used as bracing, install accord. to James Hardie Bracing Manual.

Movement Joints These are vertical and horizontal joints that control the structural movement between sheets and frame. These are required on walls over 5.4m long and at floor levels or at 3.6m MAX high whichever is the lesser. Refer to fig 3 for control joint locations and to manufacturer's specifications for further details.

Corners Internal corners require a 6mm gap between sheets, this is filled with James Hardie Joint Sealant (fig 3). External corners have either a system of continuous joining and coating around corners (fig 4) or are reinforced with a perforated angle (fig 5).

Around Joinery Fix sheets into aluminium frames. To reduce the possibility of joint cracking due to structural
movement sheet edges *should not* coincide with the sides of joinery frames. An alternative method is to provide a control joint at one side of an opening (see fig 3).

Coating Systems Must be sufficiently flexible to accommodate stresses across sheet joins and are usually 100% acrylic or pure elastomeric high build coatings. The systems must be applied by an applicator trained and approved by a reputable manufacturer. *Don't* use cement/sand mixtures.

For further details including safety instructions refer to James Hardie Harditex Installation instructions.

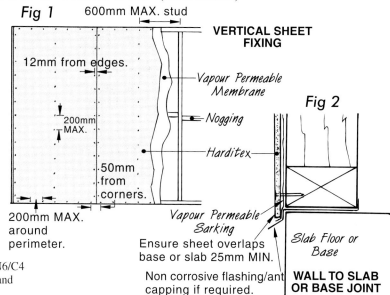

Fig 1
600mm MAX. stud
VERTICAL SHEET FIXING
12mm from edges.
Vapour Permeable Membrane
200mm MAX.
Nogging
Harditex
50mm from corners.
200mm MAX. around perimeter.

Fig 2
Vapour Permeable Sarking
Ensure sheet overlaps base or slab 25mm MIN.
Non corrosive flashing/ant capping if required.
Slab Floor or Base
WALL TO SLAB OR BASE JOINT

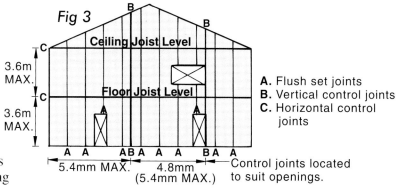

Fig 3
Ceiling Joist Level
3.6m MAX.
Floor Joist Level
3.6m MAX.
A A A B A A A B A A
5.4mm MAX. 4.8mm (5.4mm MAX.)

A. Flush set joints
B. Vertical control joints
C. Horizontal control joints

Control joints located to suit openings.
MOVEMENT CONTROL JOINTS

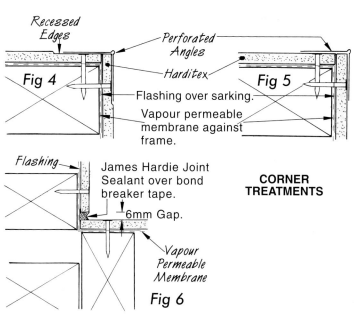

Recessed Edges
Perforated Angles
Harditex
Fig 4
Fig 5
Flashing over sarking.
Vapour permeable membrane against frame.

Flashing
James Hardie Joint Sealant over bond breaker tape.
6mm Gap.
Vapour Permeable Membrane
Fig 6

CORNER TREATMENTS

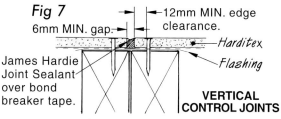

Fig 7
6mm MIN. gap.
12mm MIN. edge clearance.
Harditex
Flashing
James Hardie Joint Sealant over bond breaker tape.
VERTICAL CONTROL JOINTS

Notes: a). *Do not* apply texture directly over sealants.
b). When using textured coating see Page 145.

115

Hebel PowerPanel Installation

(For fixing to steel frames refer to Hebel instructions)
Hebel PowerPanel is attached to the timber frame as a veneer providing a 20 to 40mm cavity between. When double sided reflective foil is applied with the 40mm cavity, the winter 'R' value is approximately 2.2. The standard system has a 20mm cavity.

Hebel render or thin section render as on page 145, is applied. The following instructions are only an outline of installation. Full fixing instructions are available from the Hebel web site.

A stud frame is erected using 450mm or 600mm stud centres and aligned straight and flat. A rebate is provided to the slab perimeter as for brick veneer the rebate should allow for the panel to overhang the slab (i.e. 80mm wide rebate if using a 20mm cavity).

Steps for Construction

Step 1 Check the studs for any severe humps or hollows using a long straight edge then correct. Apply sarking as required. This is highly encouraged and is most likely required to acheive minimum wall thermal performance for the specific locality.

Step 2 Install top hat battens to the studs for attaching the panels. Battens are spaced according to specifications. Expansion Joints should be provided every 6 meters horizontally. Extreme soil conditions may require additional joints. The battens are discontinued on each side of the joint.

Step 3 Lay approved damp proof course along the rebate & up the wall and attach to the bottom plate.

Step 4 Install the first Power Panel commencing from a corner extending past the timber frame by the battern width plus 75mm in order for the return panel to meet flush. Bring into accurate plumb & level alignment for the following adjoining panel. An assistant then supports the panel while a screw is driven to hold it in position. Remaining screws are then driven. It is recommended that the external face fix method is used except where access is not possible.

Step 5 A coat of Hebel adhesive is applied to the vertical edge of the panel using the 75mm notched Hebel trowel and the second panel is erected, bedded into the adhesive then fastened.

Step 6 Provide 10mm vertical expansion joints every 6 metres. These are later filled with a foam backing rod then with polyurethane sealant.

Window sills

Step 7 At window sills the top of the Hebel panel is trimmed with a circular saw to a slope of about 15° then taken up underneath the frame. Leave a 5mm gap for frame shrinkage. Failure to do so could cause the frame to sink into the Hebel sill and deform the aluminium sill. The gap is filled with polyurethane sealant.

> **IMPORTANT NOTES**
> **a).** Be sure to bring the first panel into plumb alignment. **b).** When upper level Hebel panel walls adjoins roof, do not lay roof tiles until the walls are complete.
> **c).** When sheet bracing has been used in the frame, increase length of the panel fixing screws accordingly. **d).** Ensure you use screws with appropriate corrosion protection as nominated in the Hebel design manual.

Penetrations

All penetrations for services are neatly cut and must be sealed around with polyurethane sealant. Where panels have been cut to various lengths, the ends of reinforcing are prime painted with Fentak Reinforcing Touch Up paint.

Prepare for Coatings Fill any joints (other than expansion joints), voids and screw holes with Hebel patch, allow to dry then remove any ridges at joins using the Hebel sanding float. Fill Control Joints and gaps around windows with polyurethane sealant.

Apply Coatings See page 145

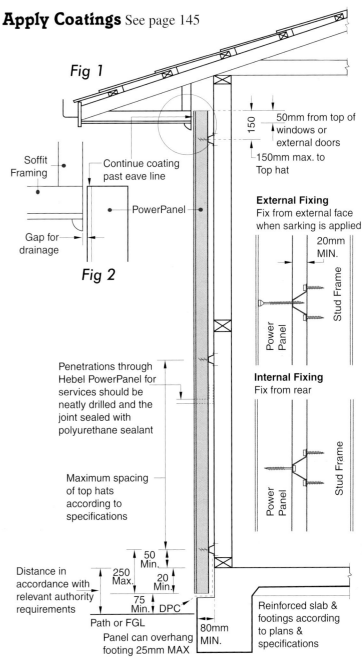

Fig 1

Soffit Framing

Continue coating past eave line

PowerPanel

Gap for drainage

Fig 2

150

50mm from top of windows or external doors
150mm max. to Top hat

External Fixing
Fix from external face when sarking is applied

20mm MIN.

Power Panel · Stud Frame

Internal Fixing
Fix from rear

Power Panel · Stud Frame

Penetrations through Hebel PowerPanel for services should be neatly drilled and the joint sealed with polyurethane sealant

Maximum spacing of top hats according to specifications

Distance in accordance with relevant authority requirements

50 Min.
250 Max.
20 Min.
75 Min. DPC
Path or FGL
Panel can overhang footing 25mm MAX
80mm MIN.

Reinforced slab & footings according to plans & specifications

116

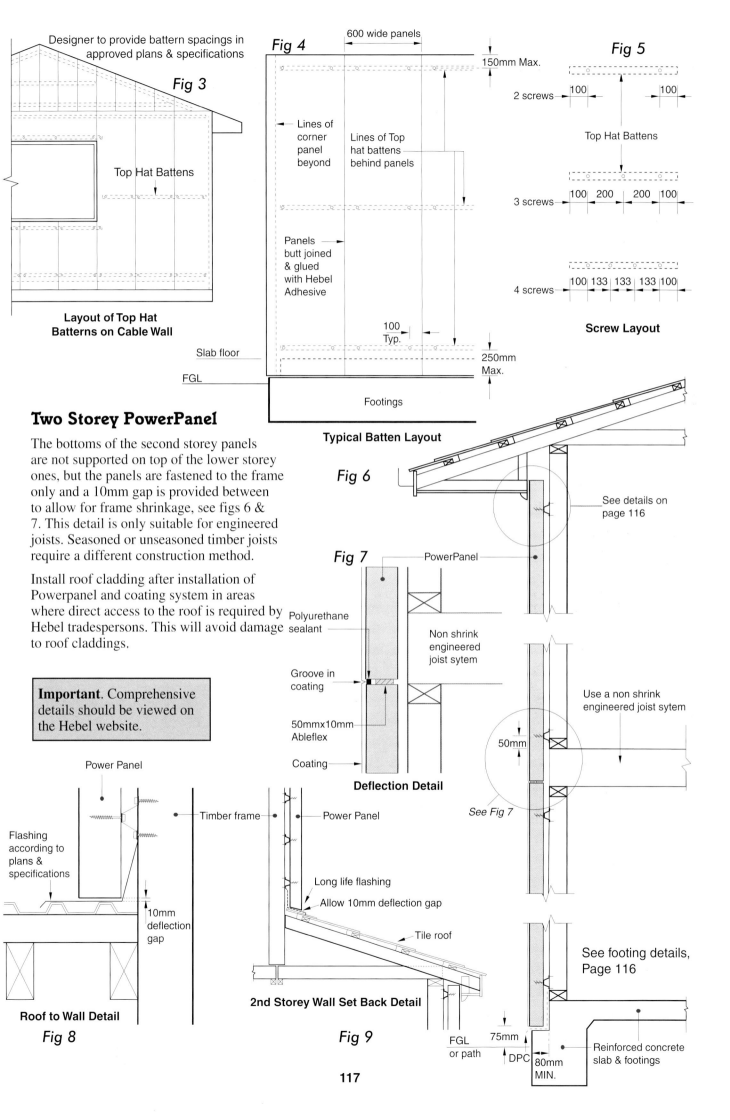

Fig 3

Designer to provide battern spacings in approved plans & specifications

Top Hat Battens

Layout of Top Hat Batterns on Cable Wall

Fig 4

600 wide panels

150mm Max.

Lines of corner panel beyond

Lines of Top hat battens behind panels

Panels butt joined & glued with Hebel Adhesive

100 Typ.

Slab floor

FGL

250mm Max.

Footings

Typical Batten Layout

Fig 5

2 screws | 100 | 100

Top Hat Battens

3 screws | 100 | 200 | 200 | 100

4 screws | 100 | 133 | 133 | 133 | 100

Screw Layout

Two Storey PowerPanel

The bottoms of the second storey panels are not supported on top of the lower storey ones, but the panels are fastened to the frame only and a 10mm gap is provided between to allow for frame shrinkage, see figs 6 & 7. This detail is only suitable for engineered joists. Seasoned or unseasoned timber joists require a different construction method.

Install roof cladding after installation of Powerpanel and coating system in areas where direct access to the roof is required by Hebel tradespersons. This will avoid damage to roof claddings.

Important. Comprehensive details should be viewed on the Hebel website.

Fig 6

See details on page 116

PowerPanel

Fig 7

Polyurethane sealant

Groove in coating

50mmx10mm Ableflex

Coating

Non shrink engineered joist sytem

Deflection Detail

Use a non shrink engineered joist sytem

50mm

See Fig 7

See footing details, Page 116

Power Panel

Flashing according to plans & specifications

10mm deflection gap

Roof to Wall Detail

Fig 8

Timber frame

Power Panel

Long life flashing

Allow 10mm deflection gap

Tile roof

2nd Storey Wall Set Back Detail

Fig 9

FGL or path

75mm

DPC

80mm MIN.

Reinforced concrete slab & footings

Installing Aluminium Windows

Ordering Aluminium Windows

Double check the rough stud opening measurements and request frames and glass to be made in accordance with the Australian Standards. Request reveals and flashings to be factory fitted.

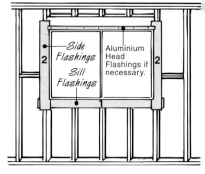

OUTSIDE VIEW WITH FLASHINGS

The window frame is inserted from the outside.

Fig 1

Receiving Delivery

a). Before arrival, prepare a safe waterproof storage area where they can be stored on their edges and secured with a cord. **Receive delivery only when you are ready to install them to prevent breakage and theft.**

b). On arrival check that each one is to size and specification.

c). Prime paint the reveals and install them as soon as possible.

Aluminium windows are easier to install if reveals are prefitted, which is normally the case. It is a good idea to have a number of wedges cut beforehand. These should be about 100mm long and taper from 6mm to nothing or use strips of DPC or hardboard. These will be used to pack the reveals plumb and level.

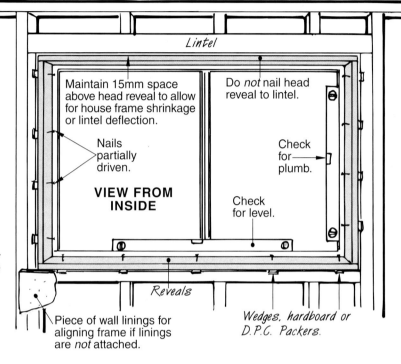

Fig 2

Maintain 15mm space above head reveal to allow for house frame shrinkage or lintel deflection.

Nails partially driven.

VIEW FROM INSIDE

Do *not* nail head reveal to lintel.

Check for plumb.

Check for level.

Piece of wall linings for aligning frame if linings are *not* attached.

Reveals

Wedges, hardboard or D.P.C. Packers.

Installing Aluminium Windows with Pre-fitted Reveals

Prior to installing, plane off half a millimetre from the back edge of the reveals to enable architraves to rest tight against the reveals. Be careful *not* to remove any wood from the inner leading edge (see fig 4).

Step 1 Place the window in the opening from the outside. Drive a temporary nail 75-100mm long into the timber head outside at an angle down and over the top of the window to prevent it from falling out.

Step 2 From the inside, pack up the sill reveal until level and straight using the wedges or packing. Ensure it is flush with internal linings. If linings have *not* been installed, hold a piece of the proposed lining on the face of the wall framing beside the sill and adjust the sill in or out until it rests flush with the lining (see fig 2). The sill can then be nailed but leave nails protruding until the sides have been straightened and plumbed.

Step 3 Repeat the above procedure for the side reveals, plumbing and wedging straight. Then adjust the frame until the reveals rest flush with the proposed wall linings. Avoid using nails, wedges or packing above the head reveal. A clear space of 15mm should be maintained for shrinkage or

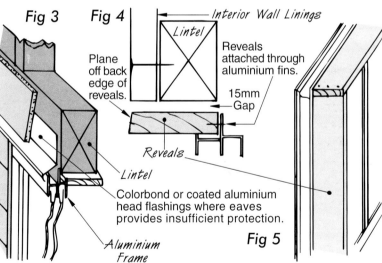

Fig 3   *Fig 4*

Lintel

Plane off back edge of reveals.

Interior Wall Linings

Reveals attached through aluminium fins.

15mm Gap

Reveals

Lintel

Colorbond or coated aluminium head flashings where eaves provides insufficient protection.

Aluminium Frame

Fig 5

deflection of the lintel. If a bow is encountered in the head reveal, this is straightened with the architrave later. When complete, the sliding or awning sash should, when closed, be parallel with the frame. This method of installation is also applicable for the installation of timber window joinery.

> *Hint:* When testing the frame for level or plumb, rest the level on the aluminium frame, not the timber reveals.

118

Installing Aluminium Sliding Doors

Installing Aluminium Sliding Doors

When resting on timber floors, the aluminium sill should be fully supported on a level timber base preferably set down in a rebate as in fig 1.

On concrete floors, the sill is also best set-down in a rebate (see fig 2).

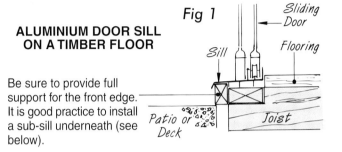

Fig 1

ALUMINIUM DOOR SILL ON A TIMBER FLOOR

Be sure to provide full support for the front edge. It is good practice to install a sub-sill underneath (see below).

The aluminium sub-sill can be supplied by some manufacturers and provides support for the sill and a much better finish to the lower front edge.

ALUMINIUM SUB-SILL

Fig 2

ALUMINIUM DOOR SILL ON A CONCRETE FLOOR

Ensure front edge of door sill is fully supported. It is good practice to install a sub-sill underneath (as above).

Step 1 Sit frame in opening already prepared.

Step 2 Before fastening sill, check jambs for plumb as the sill may require sliding sideways.

Step 3 Make sure head, jamb and sill flashings are in place and that the sill is level and fully supported through its length. Secure sill in position.

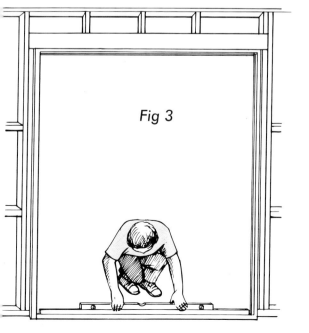

Fig 3

Step 4 Wedge jambs until they are plumb and straight, fixing top screws first. Screw fix in grooves suggested by the manufacturer or holes provided. Some manufacturers will supply aluminium frames with timber reveals already fitted to frames. In this case, nail through reveals to plumb and straighten. Screws will still be required in the aluminium frame later.

Fig 4

Step 5 Insert fixed panel by raising up into head track first and then lowering. Then push tightly into side frame. Screw fix through the frame into the fixed panel frame and fix head 'L' bracket if one is provided for securing fixed panel.

Fig 5

Step 6 Install door. Adjust the wheels until the door, when almost closed, shows a parallel vertical gap between the door and frame. Fit handle and lock following the manufacturer's instructions.

Fig 6

Interior Door Installation

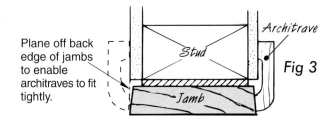

Plane off back edge of jambs to enable architraves to fit tightly.

Architrave
Stud
Jamb
Fig 3

Standard Door Sizes — 2040mm high by 870, 820, 770, 720, 620 and 520mm wide.

How to Hang Doors

Step 1 Check opening width and height.
For example, an 820mm door and using 19mm thick jambs:

Total Trim Opening Width should be:

Door	*820mm*
Jamb 19mm each side	*38*
Allowance for plumbing jambs	*10*
Clearance between door & jamb (3mm each side)	*6*
Total Trim Width	*874mm*

Total Trim Opening Height should be:

Door	*2040mm*
Head	*19*
Allowance for levelling head jamb	*3*
Allowance for floor coverings (depending on carpet & underlay, for vinyl flooring allow 12mm).	*20*
Clearance between head jamb & door	*3*
Total Trim Height	*2085mm*

Step 2 Square lines across studs to indicate the top of jambs (see fig 1). This height will equal the door height plus the 3mm gap between door top and head jamb, plus floor covering allowance of 20mm for carpet or 12mm for vinyl. After marking these heights, check the hinge side stud for plumb and straight and if necessary, tack shims of damp course or hardboard to the stud where necessary to provide a plumb surface for attaching the jamb. A straightedge and level are used for plumbing the packers (see fig 2).

Step 3 The hinge side jamb is cut to the length previously marked on the jamb stud. Then using a plane, remove the back edges from both edges of the jamb as in fig 3. The hinge positions are then marked on the jamb 250mm up from the bottom and 200mm down from the top. For extra wide or high doors, solid core doors or exterior doors use three 100mm hinges.

Step 4 Wedge the door into a door block as in fig 4 and carefully remove the back edges from both edges of the door as in fig 5. Hold the hinge jamb on the hinge edge of the door and transfer the hinge positions across to the door accurately, allowing for the head jamb plus the 3mm gap between door and head jamb. The hinges can then be fastened to the door. Use flush hinges that *do not* require rebates. Prior to installing into opening, the door can be pre-hung in its frame or hung piece-by-piece as follows.

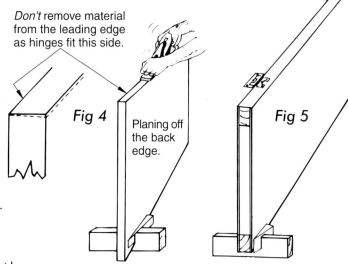

Don't remove material from the leading edge as hinges fit this side.

Fig 4

Planing off the back edge.

Fig 5

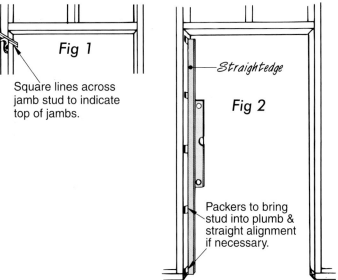

Fig 1

Square lines across jamb stud to indicate top of jambs.

Straightedge

Fig 2

Packers to bring stud into plumb & straight alignment if necessary.

Hint: Check the door for warp before hanging. Check that the lock side of the door is on the correct side before fitting hinges. The lock side has a block inside the door for installing the lock.

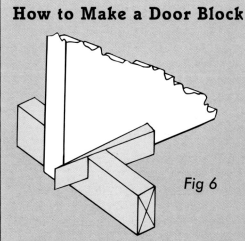

How to Make a Door Block

Fig 6

Use 100x50mm stock approx. 450mm long. Cut a slot centrally 50-75mm deep and 10-20mm wider than the thickness of the door. One side of the slot is cut square and the opposite side 6mm out of square to allow for a wedge to slide in.

Cont.

Step 5 Attach the hinge jamb onto the previously prepared jamb stud with 50 or 65mm nails (depending on jamb thickness) ensuring a flush fit with linings or slightly proud.

Step 6 The door is stood and wedged up beside the opening with the hinges aligning with the previously marked hinge positions on the jamb. The door should be in the position as in fig 7. The top screw hole is drilled and the screw fixed in place. Then fix the bottom screw in the bottom hinge. The door is then closed to check it is aligning flush with the jamb. The remaining screws are then driven. (Alternative to this, the hinge jamb can be hinged to the door prior to installation).

Step 7 The lock side jamb is cut to length. This length is found by closing the door and transferring the door height across on the jamb. The jamb length will be 3mm higher than the door. This jamb can then be packed out and nailed into position. Allow a 2-3mm gap between jamb and door.

Ensure that the jamb edges are aligning with the face of the door from top to bottom and also flush with wall linings. It is good practice to use pairs of wedges on this jamb to enable easier adjustment of the jamb.

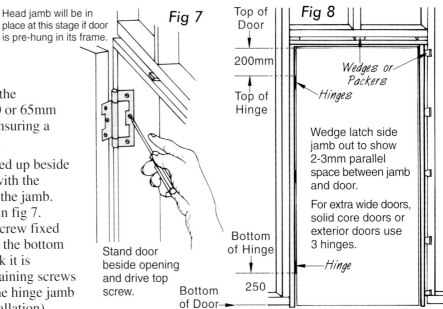

Head jamb will be in place at this stage if door is pre-hung in its frame.

Fig 7

Stand door beside opening and drive top screw.

Fig 8

Top of Door — 200mm — Top of Hinge

Bottom of Hinge — 250 — Bottom of Door

Wedges or Packers — Hinges

Wedge latch side jamb out to show 2-3mm parallel space between jamb and door.

For extra wide doors, solid core doors or exterior doors use 3 hinges.

Hinge

Where the gap between the door and jamb is too great, the wedges can be simply driven further in.

Step 8 The head jamb is cut to length and installed. Wedges are driven in each end until the head fits tightly onto the tops of the side jambs. Nails can be driven down from the top of the head jambs into the side jambs. A pair of wedges or packers are fitted centrally and the head jamb secured. The door is ready to receive door stops and latchsets or locksets.

Exterior Door Installation
(See also Pages 21 & 33)

Entries to front doors should have some form of roof covering to protect the waiting visitor from rain and the door from weather damage. If front doors are stain finished, they will require regular maintenance. Ensure all joints to the frame perimeter have been adequately sealed. Exterior doors may open in or out. Opening-out doors offer superior weather protection because the door closes outside of a rebate (see fig 2). However, these are *not* suitable for front door use.

Installation Methods
Exterior doors are hung using the same methods as for other house doors. The door as well as the frame perimeter-sides, top and bottom including end grains should be well prime painted.

Sills
Sills should be continuously supported through their length and also out close to the front edge.

On Timber Floors, sills should be fastened to the joist tops. Noggs are fitted or a full length of 90x45mm is housed into the joist tops to support the sill continuously. The house flooring is cut to fit into the back of the sill rebate (see figs 1 & 2).

On Concrete Floors, sills are installed into a rebate in the slab edge. Embed the sill into two beads of long-life sealant. Apply the beads near the front and back edges.

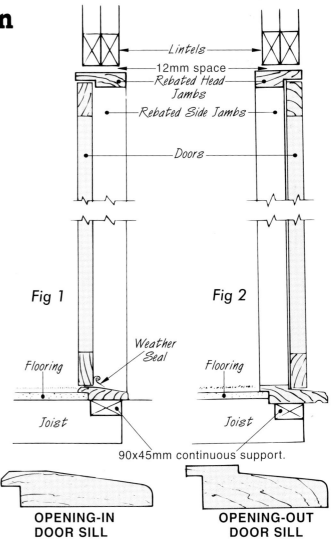

Lintels

12mm space
Rebated Head Jambs
Rebated Side Jambs

Doors

Fig 1

Fig 2

Weather Seal

Flooring

Flooring

Joist

Joist

90x45mm continuous support.

OPENING-IN DOOR SILL

OPENING-OUT DOOR SILL

Install Door Handles, Latchsets & Locksets

Step 1 Slide wedges under the door to keep it firm then square a horizontal line at the desired latch height on the door edge and continuing across both sides of the door to the handle position (see fig 1).

A cardboard template should be provided with the latch set kit. Hold this template on the line and mark the hole centres for the latch and handle mechanism.

Step 2 Drill the two holes using drill bit sizes specified on the template (see fig 2). If the latch is a mortice lock then a series of holes should be drilled, directly above one another. A chisel will be required to remove any obstructing wood which prevents a loose installation of the latch casing.

Step 3 To make the housing to accept the latch face on the door edge, install the latch, accurately mark the outer edges of the latch face on the door edge ensuring it is centrally located. Using a sharp chisel, rebate to the depth of the latch face and screw the latch in place. If the latch bolt does not move freely, further chiselling is required in the hole.

Step 4 Install the striker plate on the jamb by first closing the door and marking the latch bolt position on the jamb. Hold the striker plate on this position and mark around the bolt housing, ensuring that when the door is closed it will align flush with the jambs. Drill and chisel out the bolt housing then refit the striker plate and mark its surround. Rebate this area for the plate to fit flush. Then secure the plate. The door stops can then be attached.

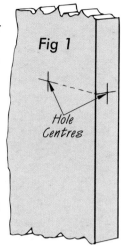

Fig 1 — Hole Centres

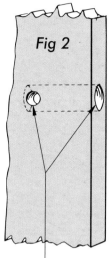

Fig 2 — Drill holes slightly larger than latch and spindle to allow room for play.

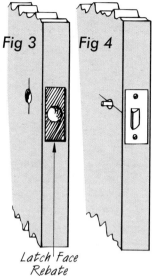

Fig 3 — Latch Face Rebate

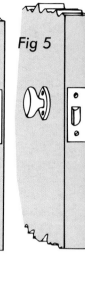

Fig 4

Fig 5

Latch housing complete ready for striker plate.

This distance should be the same as in fig 6 with the keeper held in position.

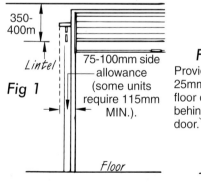

Latch / Door / Measure this distance.

Fig 6

Fig 7

Fig 7

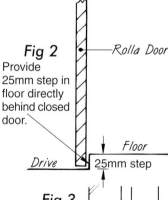

Striker Plate / Architrave / Door Stop

Fig 8

Garage Doors

Guarantees on door manufacture are usually conditional on the supplier or his agent carrying out the installation. However, rolla, tilta and overhead door manufacturers may supply installation instructions. Dressed timber jambs including a head jamb will be required.

Side jambs to support spring type tilta doors should be 45mm MIN. thick and wide enough to cross the cavity on brick veneer or double brick construction (see fig 3). Check with the manufacturer for the correct size jambs for other garage door types.

Fixing Jambs — Fix head jamb level and straight using hardboard packers or timber wedges. Then attach side jambs. Keep lower ends of side jambs just clear of the concrete to prevent wood decay and prime, paint the back and end grain of all jambs.

Fig 1

350-400mm / Lintel / 75-100mm side allowance (some units require 115mm MIN.). / Floor

Fig 2

Provide 25mm step in floor directly behind closed door. / Rolla Door / Floor / Drive / 25mm step

Rolla doors require 350-400mm MIN. clearance between head jambs and ceiling to install the roll. Refer to the manufacturer's specifications. A 25mm MIN. step should be provided between the garage door and concrete drive to prevent water penetration. The door, whether rolla or tilta, should close in front of this step.

Fig 3

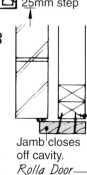

Jamb closes off cavity. Rolla Door

TOP VIEW

Ceiling Framing

Ceiling Framing Alternatives

Although timber battens have been illustrated, steel furring battens are preferred by plasterboard manufacturers and will *not* twist and warp. These faults can cause problems to the lining surfaces. However, take care to keep end-to-end joins above partitions and provide an expansion gap between. Ceilings are battened as in figs 2 & 4 or alternatively ceiling joists or trusses can be spaced at 600mm centres and 13mm Gyprock applied directly to the underside of the joists or truss bottom chords however it is better to use battens. Where trusses are spaced over 900mm apart, a ceiling joist is located between each truss as in fig 1.

Whichever method is used, support will be required around the cornice line for attaching the cornice or scocia. This can be in the form of an extra joist, batten or row of noggings.

When the upper top plates of double top plates on load bearing walls are utilised as ceiling battens, they are attached when constructing the wall. After roof construction, the remaining battens are attached. Battens are attached to the sides of top plates to support the cornice.

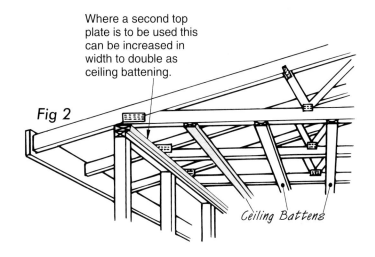

Where a second top plate is to be used this can be increased in width to double as ceiling battening.

Fig 2

Ceiling Battens

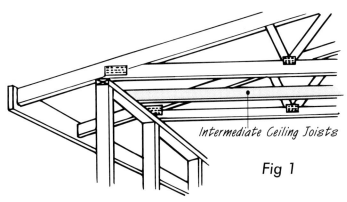

Intermediate Ceiling Joists

Fig 1

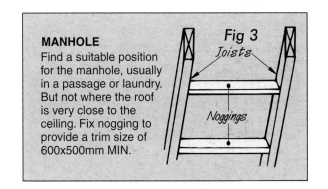

MANHOLE
Find a suitable position for the manhole, usually in a passage or laundry. But not where the roof is very close to the ceiling. Fix nogging to provide a trim size of 600x500mm MIN.

Fig 3
Joists
Noggings

Erecting Ceiling Battens

Step 1 Establish the lowest joist or truss bottom chord, add the thickness of the batten and transfer this height around the wall perimeter using a chalk line.

Step 2 Mark on the perimeter wall plates all the proposed batten positions.

Step 3 Fix both opposite side wall battens. Pack down to the chalk line using hardboard or wedges.

Step 4 Attach intermediate battens. Pack down ends to the chalk line.

To straighten the centres of intermediate battens, a string line is stretched across the battens and attached to the two opposite side wall battens previously fitted. The gauge block method of straightening is applied. Battens are packed down to the string line and secured.

Fig 4

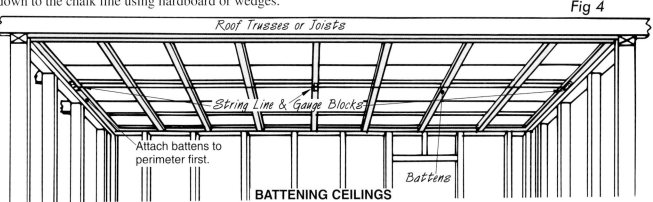

Roof Trusses or Joists

String Line & Gauge Blocks

Attach battens to perimeter first.

Battens

BATTENING CEILINGS

Gyprock® Plasterboard

(refer also to CSR's fixing instructions for further details)

Ceilings are lined before walls.

FRAMING SPACING	
Thickness (mm)	Framing Member Spacings (mm)
10 ceilings	450 standard / 600 Supa-Ceil
10 walls	600
13 ceilings	600
13 walls	600

ADHESIVE 'WALNUT' SPACING	
Board Width (mm)	Walnuts per Joist or Batten
1200	4
1350	4

RECOMMENDED FASTENINGS			
Plasterboard Thickness	Hardwood	Softwood	Metal
Gyprock® Clouts			
10mm	30mm	30mm	----
13mm	30mm	40mm	----
Gyprock® Ring Shank Nails			
10mm	25mm	30mm	----
13mm	25mm	30mm	----
Gyprock® Screws			
10mm	25mm Wood	25mm Wood	25mm
13mm	25mm Wood	32mm Wood	25mm

Installation of Ceilings

Apply walnut size daubs of stud adhesive to ceiling joists or battens at 230mm centres maximum and 200mm away from sheet edges. Attach plasterboard with paper bound edges at right angles to ceiling framing. Nail or screw recessed edges to each batten or joist and 200mm away from adhesive points. Nail ends of sheets at butt joints and around openings at 150mm centres. Nail the centres of sheets with two nails 50-75mm apart or use screws.

Important Note: 'Walnuts' of adhesive must *never* coincide with nailing points.

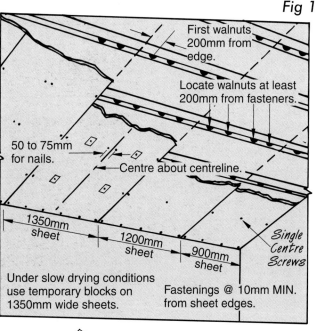

Fig 1

First walnuts 200mm from edge.
Locate walnuts at least 200mm from fasteners.
50 to 75mm for nails.
Centre about centreline.
1350mm sheet
1200mm sheet
900mm sheet
Single Centre Screws
Under slow drying conditions use temporary blocks on 1350mm wide sheets.
Fastenings @ 10mm MIN. from sheet edges.

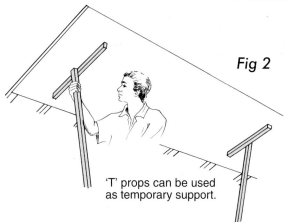

Fig 2

'T' props can be used as temporary support.

Cutting Plasterboard

Mark the sheet to length required. Hold a straight edge along the line and score through the face with a utility knife as shown in fig 3. Lift the sheet up and snap the offcut away from the scored face (see fig 4).

Then cut the back liner board (see fig 5). For double plane cuts as required around openings, make one cut with the saw, then make the intersecting cut as described above.

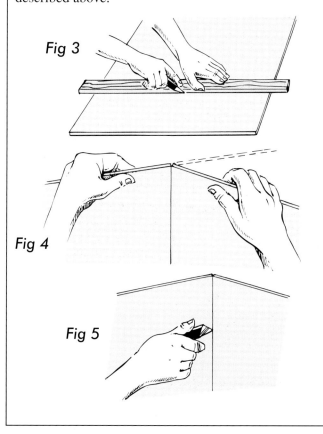

Fig 3

Fig 4

Fig 5

Backblocking (*for more detailed information refer to the manufacturer's instructions*)

Backblocking is necessary as an added precaution against cracking to ceiling joints.

Backblocking Longitudinal Ceiling Joists

Cut plasterboard backblocking panels 200mm wide and long enough to fit loosely between framing. Apply Gyprock® 'Basecoat 45/60' or cornice cement to the back blocks with a notched spreader to form 6x6mm beads @20mm ¢ at right angles to joint over the face of the block. Backblocks can be cemented in position by working above the ceiling later.

124

Backblocking End Joints in Walls & Ceilings

End or butt joints are made between framing members within 50mm of the centre. Plasterboard is cut and glued to the back of joints to fit loosely between framing as in fig 6. Apply glue the same as already described.

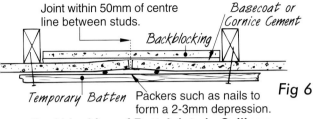

Fig 6

Backblocking of Butt Joints in Ceilings

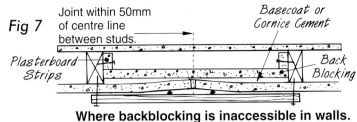

Fig 7

Where backblocking is inaccessible in walls.

Fig 8

Alternative backblocking of Butt Joints in walls

Fixing Gyprock® Plasterboard to Walls

Ensure all studs and noggings are flush with plates and that badly sprung studs have been straightened *(see Page 90)*.

Fastening

Plasterboard may be fastened to frames by nails or screws or a combination of nails and adhesive.

The latter being the most common, is described in this text. Firstly, cut the sheet to fit across the stud frame horizontally, starting with the upper sheet first. Then apply Gyprock® acrylic stud adhesive to form walnut sized blobs at approximately 300mm centres and starting 200mm in from the edges of the sheet.

Press the sheet firmly in position and nail edges and butt end joins at 150mm centres approximately 12mm in from the sheet edges. Nail at 300mm centres around openings and in corners. Drive nails just below the surface without fracturing the liner board. Hold sheets against studs for 24 hours by nails driven through temporary plasterboard blocks. Keep temporary nails away from adhesive points.

Fig 10

TEMPORARY PLASTERBOARD BLOCKS

ADHESIVE 'WALNUT' SPACING	
Board Width (mm)	Walnuts per Wall Stud
1200	3
1350	4

Fig 9

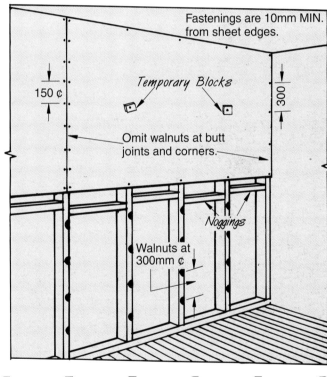

SHEET LAYOUT

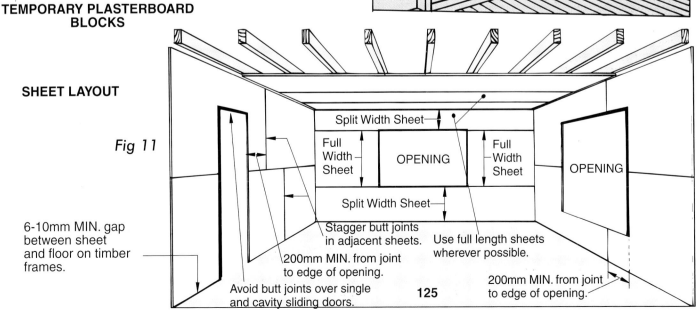

Fig 11

6-10mm MIN. gap between sheet and floor on timber frames.

Staircase Construction
(see the BCA for details).

Stairs and their proportions determine the stairwell dimensions. A minimum of two metres head height is required above steps measured vertically above nosings (see fig 1). Risers and treads must be constant throughout. A variation can cause accidents. 18 risers are the maximum permissible. Treads are required to be non-slip or have non-skid strips close to the nosings.

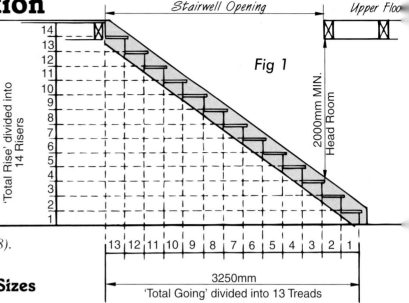

Fig 1

Timber Sizes
For stringers and treads *(see Tables, Page 128)*. Risers are usually 19mm thick.

Establish Tread (Going) & Riser Sizes

Step 1 Establish tread and riser dimensions by dividing the floor to floor height (total rise) into a suitable number of risers each rise being between 150 & 190mm. Then calculate the tread width in proportion to the rise according to *'ideal stair proportions'*. Adjust both measurements to conform to the 'Total Rise' and 'Total Going'.

Treads are usually between 250 and 300mm wide, however the BCA permits 240 MIN. to 355mm MAX. Risers are usually between 150-190mm in height, while the BCA permits 115 MIN. to 190mm MAX. The triangle created by the 'Total Rise' and the 'Total Going' can be drawn on the floor and all measurements established there.

EXAMPLE CALCULATION FROM FIG 1
Total Going = 3250 ÷ 13 Treads = 250mm Treads
Total Rise   = 2450 ÷ 14 Risers = 175mm Risers

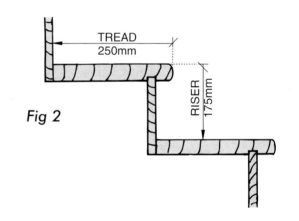

Fig 2

Ideal Stair Proportions
The following calculation can be carried out to find the correct proportions. *'The going plus twice the riser should equal between 575-650mm'*. However the BCA permits a range between 550-700mm.

Note: The space between open risers *should not* permit a sphere of 125mm to pass through.

Marking Out

Step 2 To mark out the riser and tread faces, firstly mark the nosing line (see fig 3). This is approx. 40mm below the top edge of the stringers. Then the 'Setting-out-line' is marked (see fig 4).

This is found by marking the 'going' and 'rise' on the rafter square and holding these points on the nosing line then measuring square off the intersection where the square meets the nosing line, fig 3.

The 'Setting-out-line' is gauged through this point as in fig 4. A fence or steel guides are attached to the square as in fig 5. Riser height and tread depth marks on the square must intersect the new setting out line. The fence enables tread and riser lines to be repeated accurately.

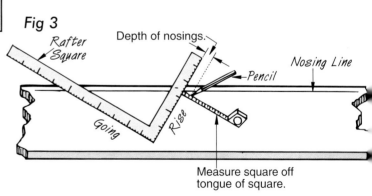

Fig 3

Measure square off tongue of square.

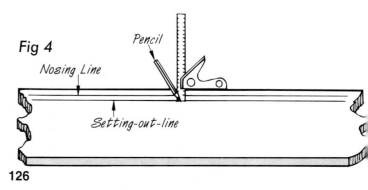

Fig 4

Cont.

Step 3 The faces of treads and risers are then marked by sliding the square along the stringer (see fig 5). Accuracy in the placement of the square on each previously marked riser and tread intersection is important and a sharp pencil is essential.

The marked out stringer can be checked by laying the stringer on the triangle previously marked on the floor.

Step 4 After marking the tops and faces of treads and risers, mark their back edges allowing for the tapering wedge to be inserted.

This is carried out by using tread and riser templates as a guide (see fig 6). The templates are cut out of ply or masonite and must include the thickness of the tread or riser whichever is being cut plus the thickness and shape of the tapering wedge. It is best to cut the wedges prior to making these templates to guarantee the correct taper is allowed for. All wedges must be of identical shape. They are best cut on a table saw using a jig. Local joinery shops will usually cut them inexpensively.

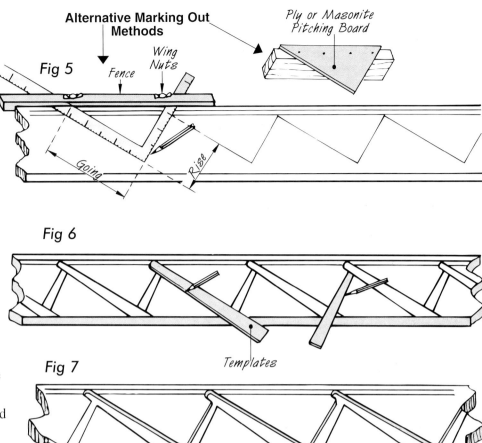

Alternative Marking Out Methods

Fig 5

Fence · *Wing Nuts* · *Ply or Masonite Pitching Board* · *Going* · *Rise*

Fig 6

Templates

Fig 7

Rebate housings 12mm deep.

When marking the shape of the tread template, include the nosing. The tread template is laid up to the nosing line along the previously marked tread line. The riser template is laid on the previously marked riser line. Repeat these four steps on the other stringer.

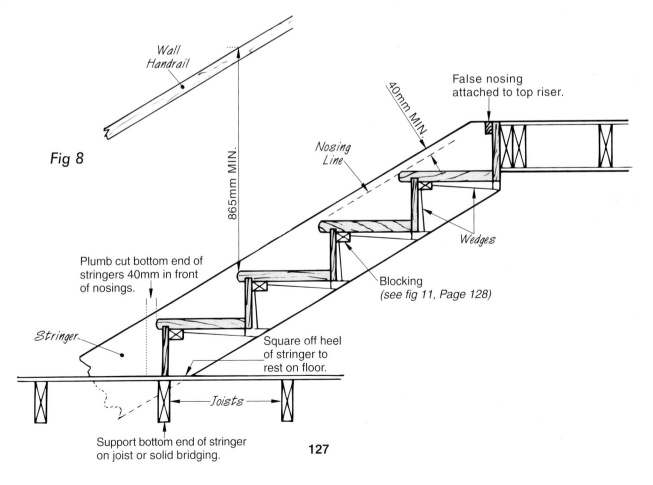

Fig 8

Wall Handrail

865mm MIN.

40mm MIN.

False nosing attached to top riser.

Nosing Line

Wedges

Plumb cut bottom end of stringers 40mm in front of nosings.

Blocking (see fig 11, Page 128)

Stringer

Square off heel of stringer to rest on floor.

Joists

Support bottom end of stringer on joist or solid bridging.

127

Cutting Out and Fitting

Step 5 Housings can be cut out using either a saw and chisel or a router. Cut housings to a depth of 12mm (see fig 7).

Step 6 Cut treads and risers to length. Assemble the staircase by first installing the top and bottom treads, then the remaining treads. Risers are then fitted, wedging and gluing their ends into position. Install corner blocks or battens for fastening the top edges (see figs 10 & 11). Stairs containing closed risers as illustrated, can be constructed without wedges. However without them, squeaking and creaking stairs are inevitable.

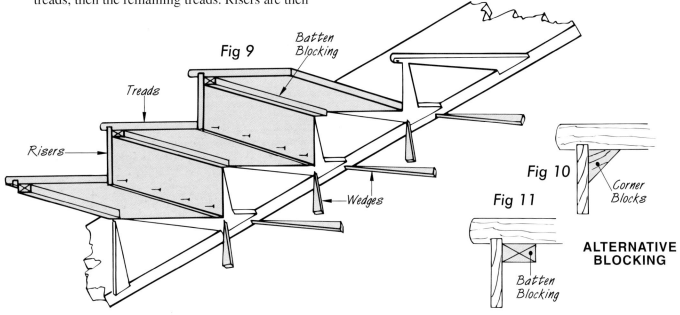

Fig 9

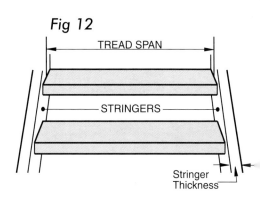

Fig 10

Fig 11

STAIR TREAD THICKNESS Minimum Tread Width 240mm

TREAD SPAN	UNSEASONED						SEASONED						
	F5	F7	F8	F11	F14	F17	F5	F7	F8	F11	F14	F17	F22
900	50	50	50	38	38	38	45	45	35	35	35	35	35
1000	50	50	50	50	50	38	45	45	45	45	35	35	35
1100	50	50	50	50	50	50	70	45	45	45	45	45	35
1200	75	50	50	50	50	50	70	70	70	45	45	45	45
1300	----	75	75	75	50	50	70	70	70	70	45	45	45
1400	----	75	75	75	75	50	70	70	70	70	70	70	45
1500	----	75	75	75	75	75	70	70	70	70	70	70	70

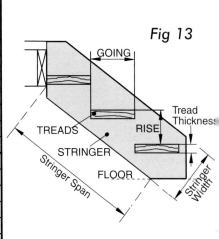

Fig 12

Fig 13

STAIR STRINGER SIZES

Tables kindly provided by 'Timber Queensland'. Suitable for all states.

STRINGER SPAN	TREAD SPAN	UNSEASONED Min. Stringer Thickness = 50mm Max. Housing Depth = 15mm						SEASONED Min. Stringer Thickness = 50mm Max. Housing Depth = 15mm						
		F5	F7	F8	F11	F14	F17	F5	F7	F8	F11	F14	F17	F22
UP TO 3600mm	900	225	225	225	225	225	225	220	220	220	220	220	220	220
	1000	250	225	225	225	225	225	240	220	220	220	220	220	220
	1100	250	225	225	225	225	225	240	220	220	220	220	220	220
	1200	250	225	225	225	225	225	240	240	220	220	220	220	220
	1300	250	225	225	225	225	225	----	240	240	220	220	220	220
	1400	275	250	250	225	225	225	----	240	240	220	220	220	220
	1500	275	250	250	250	225	225	----	----	240	240	220	220	220
3600mm to 4300mm	900	275	250	250	225	225	225	----	240	240	220	220	220	220
	1000	275	275	250	250	225	225	----	----	240	240	220	220	220
	1100	275	275	250	250	250	225	----	----	----	240	240	220	220
	1200	300	275	275	250	250	225	----	----	----	240	240	220	220
	1300	300	300	275	275	250	250	----	----	----	----	240	240	220
	1400	300	300	275	275	250	250	----	----	----	----	----	240	220
	1500	----	300	300	275	275	250	----	----	----	----	----	240	220

Constructing Exterior Timber Stairs

Step 1 Working from a pair of sawstools, mark on both stringers the tread and riser outline as in fig 1. To obtain tread and riser sizes *(see Page 128)*.

Cut tread housing trenches if required as in fig 3 or cut step shaped stringers (see fig 2). Cut ends of stringers to shape. Prime paint all raw ends with oil based wood primer. Hold one stringer on its proposed position on the landing and resting on the lower end on the ground. Mark the footing hole positions. Attach treads to stringers fastening the top and bottom ones first.

Step 2 Excavate footing holes. Bolt galvanised anchoring brackets (if required) to the stringers.

Step 3 Bolt steps in position and support their lower ends on temporary props or blocks. Construct formwork for the concrete pad at the lower end and pour concrete to footing holes and pad. Allow concrete to harden 48 hours or more before removing supports.

Note: Stringer sizes in table on Page 128 are *not* suitable for step shaped stringers as in fig 2.

Step Shaped Stringers

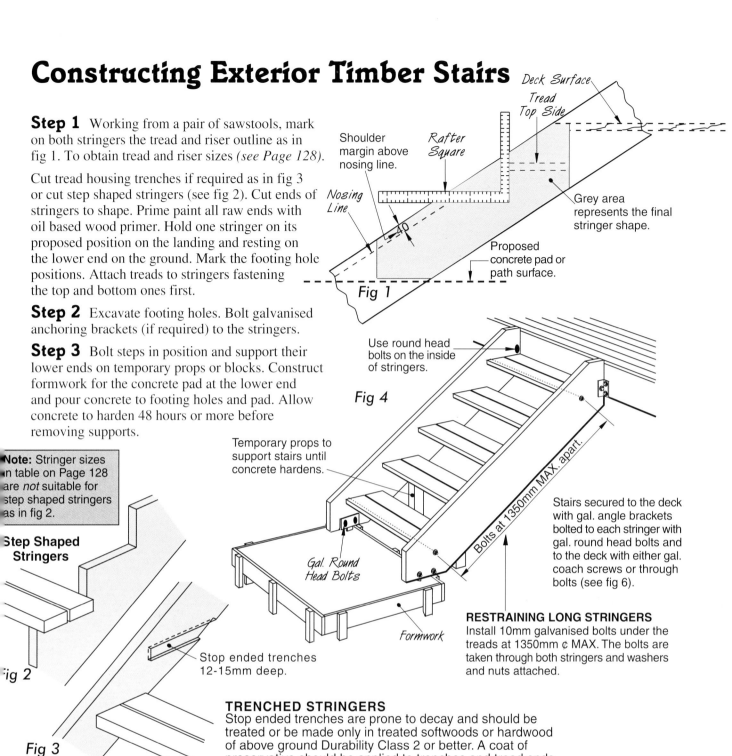

Fig 1

Deck Surface
Tread Top Side
Rafter Square
Shoulder margin above nosing line.
Nosing Line
40
Proposed concrete pad or path surface.
Grey area represents the final stringer shape.

Fig 4

Use round head bolts on the inside of stringers.

Temporary props to support stairs until concrete hardens.

Gal. Round Head Bolts

Formwork

Bolts at 1350mm MAX. apart.

Stairs secured to the deck with gal. angle brackets bolted to each stringer with gal. round head bolts and to the deck with either gal. coach screws or through bolts (see fig 6).

RESTRAINING LONG STRINGERS
Install 10mm galvanised bolts under the treads at 1350mm ¢ MAX. The bolts are taken through both stringers and washers and nuts attached.

Fig 2

Stop ended trenches 12-15mm deep.

Fig 3

TRENCHED STRINGERS
Stop ended trenches are prone to decay and should be treated or be made only in treated softwoods or hardwood of above ground Durability Class 2 or better. A coat of preservative should be applied to trenches and tread ends.

Anchoring Stairs to Decks

Stairs can be adjoined to the deck by the stringers and secured by fastenings through the stringers into the deck joists as in fig 5 or by galvanised angle brackets as in figs 4 & 6. Housing outdoor stringers is *not* advisable as they are prone to decay. Steel stringers are fastened to the deck with bolts through flat plates welded to the stringer ends as in fig 7.

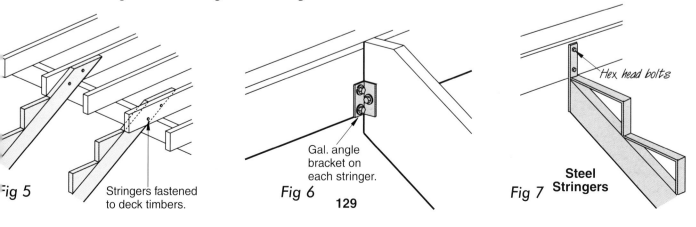

Fig 5

Stringers fastened to deck timbers.

Fig 6

Gal. angle bracket on each stringer.

Fig 7

Steel Stringers

Hex head bolts

Anchoring Stringers at the Bottom Ends

Timber stringers can be anchored at the bottom ends by fabricated galvanised brackets bolted to each stringer as in figs 8 & 9. The bracket is embedded in concrete. Drill bolt holes through bracket before galvanising. Keep underside of brackets clear of the concrete pad or path by 12-20mm.

Steel Stringers (use hot dip galvanised)

Steel stringers have their lower ends embedded in concrete or sometimes bolted to a concrete pad. To prevent corrosion, take care that all honeycombing is removed around embedded stringers and that the concrete surface slopes away from the stringers. Ensure moisture *can not* enter hollow stringers.

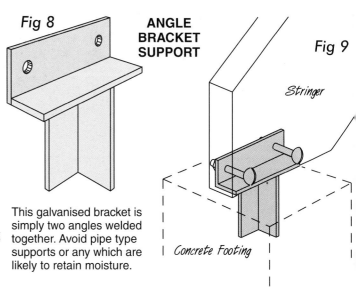

Fig 8

ANGLE BRACKET SUPPORT

Fig 9

This galvanised bracket is simply two angles welded together. Avoid pipe type supports or any which are likely to retain moisture.

Concrete Stair Construction

(For selecting handrail sizes see Tables on Page 133)

Concrete stairs including the reinforcement, its placement and footings should be designed by an Engineer.

Preparing Formwork

Step 1 Commence by laying bearers, joists and decking.

Step 2 Erect stringers and riser formwork. Fit a carriage riser joist to prevent distortion of the risers when the concrete is poured.

Step 3 Insert reinforcing mesh from the top and position according to plans and specifications. Then pour concrete. Use a vibrator to remove all trapped air.

Precast Concrete Stairs

Precast stairs are available. The footing and seating for the top riser are prepared and the stairs lifted into position by crane.

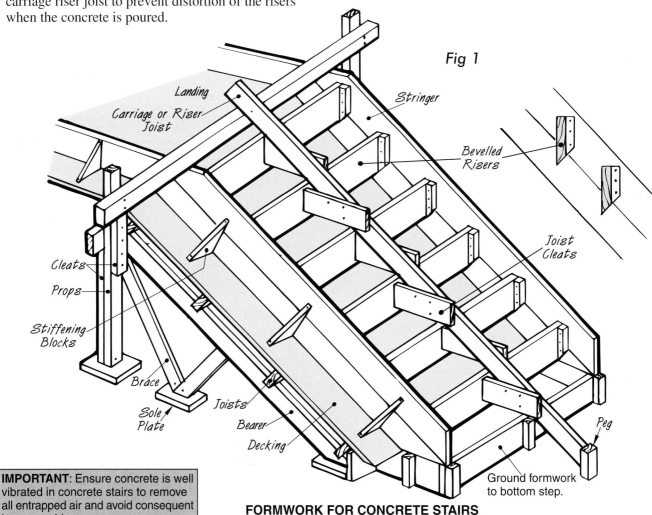

Fig 1

FORMWORK FOR CONCRETE STAIRS

IMPORTANT: Ensure concrete is well vibrated in concrete stairs to remove all entrapped air and avoid consequent honeycombing.

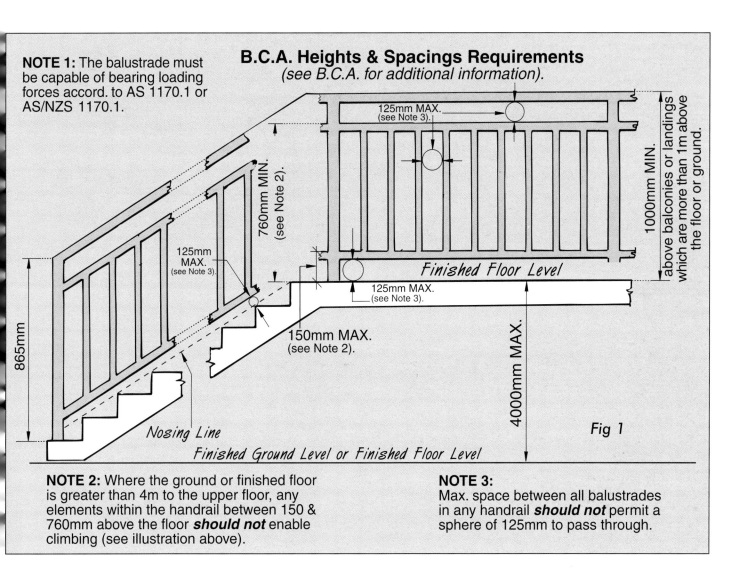

NOTE 1: The balustrade must be capable of bearing loading forces accord. to AS 1170.1 or AS/NZS 1170.1.

B.C.A. Heights & Spacings Requirements
(see B.C.A. for additional information).

125mm MAX. (see Note 3).

1000mm MIN. above balconies or landings which are more than 1m above the floor or ground.

760mm MIN. (see Note 2).

125mm MAX. (see Note 3).

Finished Floor Level

125mm MAX. (see Note 3).

865mm

125mm MAX. (see Note 3).

150mm MAX. (see Note 2).

4000mm MAX.

Fig 1

Nosing Line

Finished Ground Level or Finished Floor Level

NOTE 2: Where the ground or finished floor is greater than 4m to the upper floor, any elements within the handrail between 150 & 760mm above the floor *should not* enable climbing (see illustration above).

NOTE 3: Max. space between all balustrades in any handrail *should not* permit a sphere of 125mm to pass through.

Handrailing
(For selecting handrail sizes see the Tables on Page 133)

Important Note: Handrail design and construction is critical as litigation for injury or even death resulting from poor construction could confront the builder even years after construction. Many common designs and methods of construction are suspect. In fact only a few could be considered safe long term. Weakening of the structure through wood decay and corrosion of fasteners is a common fault.

Near the coast you will need better coatings than hot dip gal. and in all situations handrail fastenings must *never* fail. Look for deck designs that have built-in permanent handrail reliability such as in figs 2&3 & 5-8.

Design Requirements
The handrail heights and spacing requirements in the BCA are illustrated in fig 1. Check with the Local Council Building Depart. for any more recent changes to these requirements nationally or locally before adopting them.

Wire (Wire Rope) Balustrade
See the BCA for extensive details.

What Timber to Use
For above ground use H3 ACQ or Copper Azole treated softwood or better or hardwood of above ground durability Class 1 or 2. Seasoned will perform better in exposed situations. *Species suggested are:* Jarrah, Ironbark, Spotted Gum, Kwila (Merbau).

Where posts are used in-ground but their above ground portions are accessible to human contact use H5 ACQ or Copper Azole treated softwood or hardwood of inground Durability Class 1 or 2 *See Page 8 for further information on timber selection and treatment.*

Handrail Construction
Size of Handrail Posts
Handrail posts must *not* be smaller in section than 70x70mm F8 or 90x45mm F8 for post spacing up to 2000mm and spacings over 2000mm must have the posts and fastenings designed by a structural engineer.

Note: CCA treated timber is *not* permitted on decks, handrails or steps or in any human contact locations, see also Page 8.

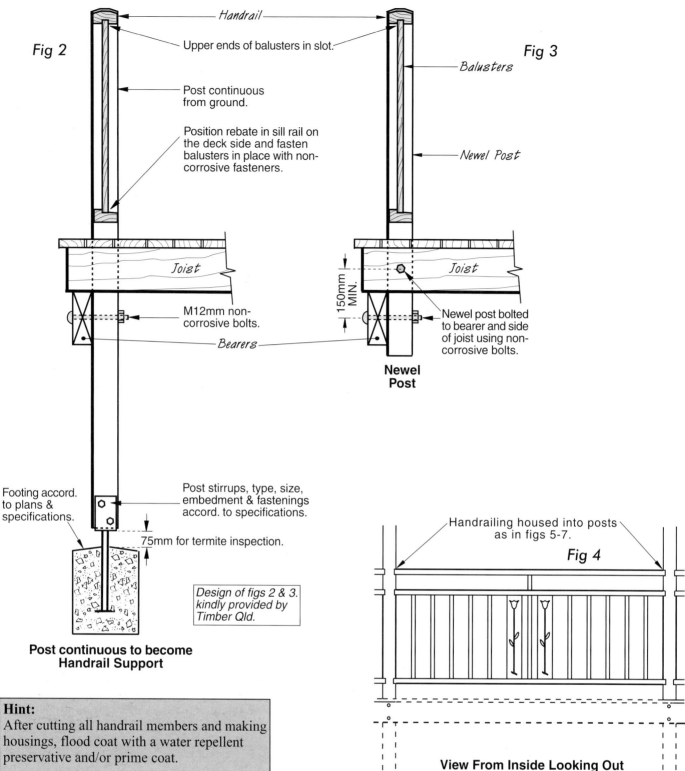

Fig 2

Handrail

Upper ends of balusters in slot.

Post continuous from ground.

Position rebate in sill rail on the deck side and fasten balusters in place with non-corrosive fasteners.

Joist

M12mm non-corrosive bolts.

Bearers

Footing accord. to plans & specifications.

Post stirrups, type, size, embedment & fastenings accord. to specifications.

75mm for termite inspection.

Design of figs 2 & 3. kindly provided by Timber Qld.

Post continuous to become Handrail Support

Fig 3

Balusters

Newel Post

Joist

150mm MIN.

Newel post bolted to bearer and side of joist using non-corrosive bolts.

Newel Post

Hint:
After cutting all handrail members and making housings, flood coat with a water repellent preservative and/or prime coat.

Fig 4

Handrailing housed into posts as in figs 5-7.

View From Inside Looking Out

Fastening Handrails to Posts

Handrails can be cut to fit between posts as in figs 2, 3 & 4. It is essential when applying this method that the joints are properly designed. Fig 5 is a suggested method. Figs 6&7 are suggested fastenings. Fig 8 offers an alternative rail to post joint. It has the rail bolted to the posts. These could be face mounted or checked in 10 or 12mm. The post may terminate at rail height or continue through to become pergola posts.

Note: Butt joints with skew nailing should *never* be applied to rail-to-post joints.

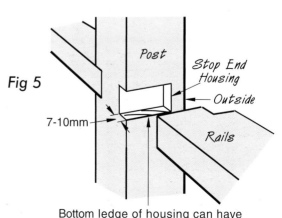

Fig 5

Post

Stop End Housing

Outside

Rails

7-10mm

Bottom ledge of housing can have a slope for moisture to drain off.

Cont.

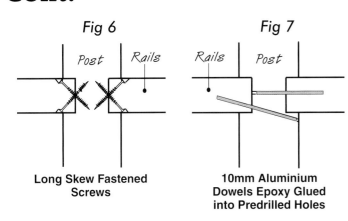

Fig 6
Post   Rails

Long Skew Fastened Screws

Fig 7
Rails   Post

10mm Aluminium Dowels Epoxy Glued into Predrilled Holes

Another Method
Deck posts continue above deck to become handrail posts and sometimes also continue to become pergola posts. Vertical balusters can be screw fastened.

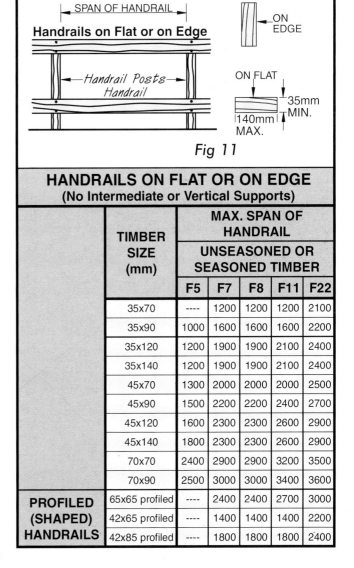

Fig 8

IMPORTANT: Use corrosion proof fastenings throughout. Handrail posts should be checked from time to time for deterioration through corrosion and wood decay.

Fastening Balusters
Refer to your plans or specifications for the correct fastener. In figs 2 & 3 the upper ends of the balusters are fastened in a slot and the lower ends into a rebate. *Don't* use a slot in the lower rail as the ends of the balusters *can't* dry out as well after wet weather even if holes are drilled between them. For all handrails use H3 treated, seasoned softwood or seasoned hardwood of above ground Durability Class 1 or 2.

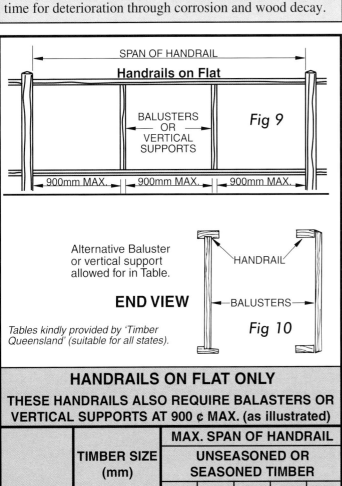

SPAN OF HANDRAIL
Handrails on Flat

BALUSTERS OR VERTICAL SUPPORTS

Fig 9

900mm MAX.   900mm MAX.   900mm MAX.

Alternative Baluster or vertical support allowed for in Table.

HANDRAIL

END VIEW

BALUSTERS

Fig 10

Tables kindly provided by 'Timber Queensland' (suitable for all states).

HANDRAILS ON FLAT ONLY
THESE HANDRAILS ALSO REQUIRE BALASTERS OR VERTICAL SUPPORTS AT 900 ¢ MAX. (as illustrated)

TIMBER SIZE (mm)	MAX. SPAN OF HANDRAIL UNSEASONED OR SEASONED TIMBER				
	F5	F7	F8	F11	F22
35x70	1700	2400	2400	2400	3000
35x90	2300	2500	2900	3200	3600
35x120	2900	3300	3600	3600	3600
35x140	3200	3400	3600	3600	3600
45x70	1900	2200	2600	2800	3200
45x90	2700	2900	3100	3400	3600
45x120	3200	3400	3600	3600	3600
70x70	----	2900	2900	3200	3500
70x90	----	3200	3400	3600	3600
PROFILED (SHAPED) HANDRAILS 65x65 profiled	----	2400	2400	2700	3000
42x65 profiled	----	2000	2000	2000	2700
42x85 profiled	z---	2700	2700	3000	3400

SPAN OF HANDRAIL
Handrails on Flat or on Edge

Handrail Posts
Handrail

ON EDGE

ON FLAT
140mm MAX.   35mm MIN.

Fig 11

HANDRAILS ON FLAT OR ON EDGE
(No Intermediate or Vertical Supports)

TIMBER SIZE (mm)	MAX. SPAN OF HANDRAIL UNSEASONED OR SEASONED TIMBER				
	F5	F7	F8	F11	F22
35x70	----	1200	1200	1200	2100
35x90	1000	1600	1600	1600	2200
35x120	1200	1900	1900	2100	2400
35x140	1200	1900	1900	2100	2400
45x70	1300	2000	2000	2000	2500
45x90	1500	2200	2200	2400	2700
45x120	1600	2300	2300	2600	2900
45x140	1800	2300	2300	2600	2900
70x70	2400	2900	2900	3200	3500
70x90	2500	3000	3000	3400	3600
PROFILED (SHAPED) HANDRAILS 65x65 profiled	----	2400	2400	2700	3000
42x65 profiled	----	1400	1400	1400	2200
42x85 profiled	----	1800	1800	1800	2400

Architraves

Door & Window Architraves

Architraves are mitre cut and mitre butt joined, glued together and nailed to wall framing on one side, and door jambs or window reveals on their opposite side.

Fitting Architraves

Ensure linings are flush or slightly back from jambs or reveals.

Step 1 Cut a 45° mitre on one end of architrave (1), see figs 1 & 2. Hold it in position with the bottom corner of the mitre 2-3mm past the opening. This provides a 2-3mm parallel quirk (border) around the opening (see figs 4 & 5). Mark the opposite end, again allowing the quirk. Cut and tack this architrave in position. Fix each end first then tack the centres maintaining the quirk. With a little practice, this quirk is maintained by eye.

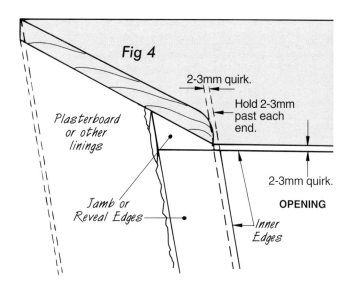

Fig 4

2-3mm quirk.

Hold 2-3mm past each end.

Plasterboard or other linings

2-3mm quirk.

OPENING

Jamb or Reveal Edges

Inner Edges

Step 2 Mitre the top end of architrave (2) and hold this architrave in position mitre-to-mitre. On doors, architraves (2 & 3) have their bottom end square cut first. Mark and cut the opposite end and tack in position. Leave nail heads protruding. Remember to glue the mitre joint (use PVA). Following this procedure, fit the remaining architraves. When encountering a mitre that is not tight, lightly dress the loose mitre until it fits, slightly backing off (undercutting) the end grain using a plane. When mouldings are to be clear or stain finished, clean off excess PVA with a damp cloth otherwise a permanent stain will result. After all mitres are tight, all nail heads can be punched.

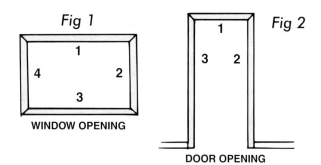

Fig 1

1
4   2
3

WINDOW OPENING

Fig 2

1
3   2

DOOR OPENING

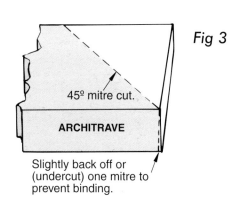

Fig 3

45° mitre cut.

ARCHITRAVE

Slightly back off or (undercut) one mitre to prevent binding.

Plane off back edge of reveals and jambs to enable architraves to fit tightly against reveals and wallboard.

Architraves are fixed with 40x2.0mm bullet head nails and punched below the surface.

Fig 5

Jamb or Reveal

Wedges

Studs

2-3mm quirk.

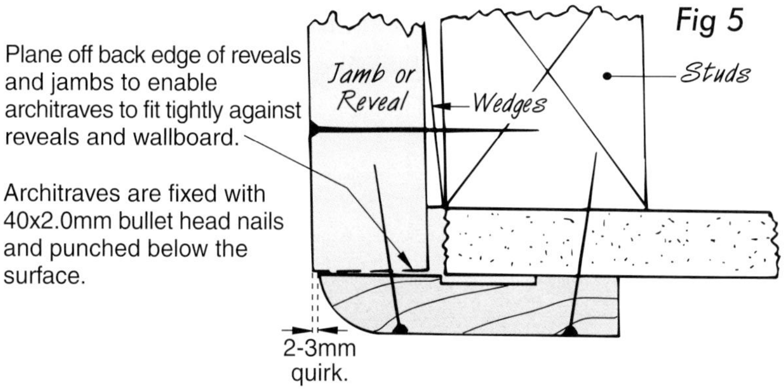

Awkward Situations

Where door and window openings are close to walls, architraves are ripped down to fit the narrow space (see fig 6). Where this space is 10-20mm wider than the normal architrave, a wider architrave is used and dressed to fit (see fig 7). This simplifies painting or wallpapering and looks tidier. A narrow side architrave adjoining a normal head architrave will necessitate square cutting part of the end of the head architrave (see fig 6). A wide side architrave adjoining a normal head architrave will necessitate square cutting part of the top of the side architrave (see fig 7).

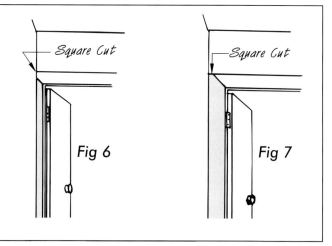

Square Cut

Square Cut

Fig 6

Fig 7

Skirtings

Internal corners are *not* mitred but scribe cut to fit the contour of the previously fitted mouldings.

Cutting & Fitting Skirtings

To obtain skirting lengths, use a tape measure or measuring sticks.

Step 1 The first moulding (1), as in fig 1 is square cut both ends to fit tightly between walls.

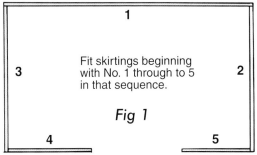

Fit skirtings beginning with No. 1 through to 5 in that sequence.

Fig 1

Order of Fitting Skirtings

Step 2 Skirting (2), fig 1 is cut to fit the contour of the first moulding as in figs 2 & 4. First cut a mitre as in fig 5. Then with a coping saw, remove all the end grain wood slightly undercutting as in fig 6. Cut the skirting to length, applying a square cut to the opposite end. Ensure a tight fit.

To transfer the contour of the first board on large skirtings, it may be necessary to cut a short length as above and adjust the cut until it fits. Use this as a pattern. When nailing skirtings, ensure the top edge is nailed to the studs and the lower edge to the bottom plate.

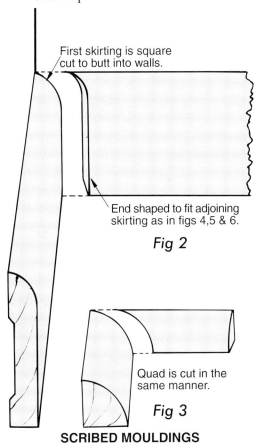

First skirting is square cut to butt into walls.

End shaped to fit adjoining skirting as in figs 4,5 & 6.

Fig 2

Quad is cut in the same manner.

Fig 3

SCRIBED MOULDINGS

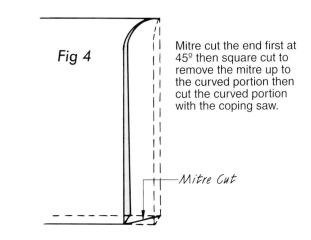

Fig 4

Mitre cut the end first at 45º then square cut to remove the mitre up to the curved portion then cut the curved portion with the coping saw.

— *Mitre Cut*

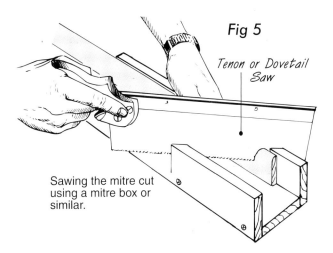

Fig 5

Tenon or Dovetail Saw

Sawing the mitre cut using a mitre box or similar.

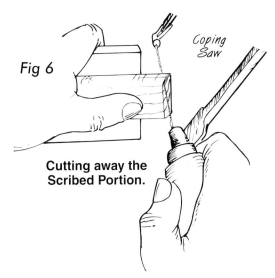

Fig 6

Coping Saw

Cutting away the Scribed Portion.

Step 3 Continue fitting the remaining skirtings as described. External corners are mitre joined.

END-TO-END JOINS

Fig 7

End-to-end joins in a length of skirting or scotia are mitre cut, *not* butt jointed.

Wet Areas Requirements

IMPORTANT: a). The author does *not* recommend the use of plasterboard (Gyprock) Gypsum-based linings in wet areas. For even if the surface can be made *waterproof* by the application of other products, the joints can, in time, became vulnerable, as sealants *do* commonly separate from adjoining materials.
b). *Waterproof* & **Water resistant** are *not* the same. *Waterproof* is impervious — **Water resistant** is *not.*

Wet areas are those where water is supplied by a water supply system and includes: bathrooms, showers, laundries, WC or sanitary areas but does *not* include *kitchens* and *bars.*

Falls to waste in shower recesses should be between 1:60 and 1:80 — all other floors between 1:80 and 1:100.

Surfaces must be free of dips or hollows to prevent ponding.

Waterproofing Materials for Floors

Liquid membranes, and PVC sheet membranes must be applied by licenced applicators. This work must comply with AS 3740 'Waterproofing of wet areas in Residential Buildings'.

Waterproofing as Recommended in the BCA

Where to Apply Waterproofing

Enclosed Showers:-

(a). Without a Hob: *Waterproof* the whole shower floor to the screen bottom edge or water stop and walls up to 150mm above the substrate of the shower floor.

Walls above this height up to 1800mm above floor surface are to be **water resistant**. All junctions internal, external and horizontal in the recess are to be *waterproof* to 40mm each side of the junction and to 1800mm high.

(b). With a Hob: *Waterproof* the whole shower floor and hob and extend waterproofing up the walls 150mm above the shower floor substrate or to 25mm above the maximum possible water retainment level whichever being the greater. Walls above this height must be **water resistant**. All junctions, internal, external and horizontal within the recess are to be *waterproof* up to 40mm each side of the junction and to 1800mm high.

(c). With a Step Down: *Waterproof* the whole shower floor and step down. Extend waterproofing up the walls 150mm above the substrate of the shower floor or to 25mm above the maximum possible water retainment level whichever being the greater. Walls above this height must be **water resistant**. All junctions, internal, external and horizontal within the recess are to be *waterproof* up to 40mm each side of the junction and to 1800mm high.

(d). With a Preformed Base: Ensure perimeter of waste is thoroughly sealed. Walls above base are to be **water resistant** all junctions, internal, external and horizontal within the recess are to be *waterproof* up to 40mm each side of the junction.

Baths & Spas:-

Waterproof all penetrations, pipes, spouts, taps and wastes etc. The entire floor should be *waterproofed* when constructed of timber, plywood, particleboard or wood based products. When constructed of concrete or compressed fibre cement or Scyon wet area flooring they need only be **water resistant**. Walls around baths or spas are to be **water resistant** to 150mm MIN. above the top edges of the vessel as well as the exposed surfaces around the vessel from its top edge to the floor. *Waterproof* the edges where baths and spas adjoin walls, the sides of the bath where the lip rests on or overlaps horizontal or vertical panels and the floor junction of any panels.

Showers without Enclosures:-

All penetrations, pipes, wastes, etc must be *waterproof*. *Waterproof* the entire floor when it is constructed from timber, plywood, particleboard or any wood product. When constructed from concrete or compressed fibre cement or Scyon wet area flooring, *waterproof* out to a radius of 1500mm MIN. from the shower connection at the wall (see figs 1 and 3). All other or remaining surfaces are to be **water resistant**.

Showers above Baths:-

See figs 4 and 5. For walls and penetrations see requirements for showers.

Inserted or Sunken Baths:-

Waterproof the entire shelf area including the water stop beneath the bath lip and extend 5mm MIN. above the tiled surface. *Waterproof* up the walls 150mm MIN. above the bath lip.

Walls adjoining Basins, Laundry Tubs and Sinks etc:-

Not necessary to *waterproof* floors. When the vessel is located within 75mm of the walls, the walls should be **water resistant** to 150mm MIN. above the vessel. Vessels fixed directly to the wall should have all adjoining edges *waterproofed*.

WC's and Laundries:-

The whole floor area is to be **water resistant**. All wall to floor junctions are *waterproofed* to 25mm MIN. above finished floor surface and sealed to the floor. When using flashings, the horizontal leg must be 40mm MIN. wide.

Waterproofing of Unenclosed Showers

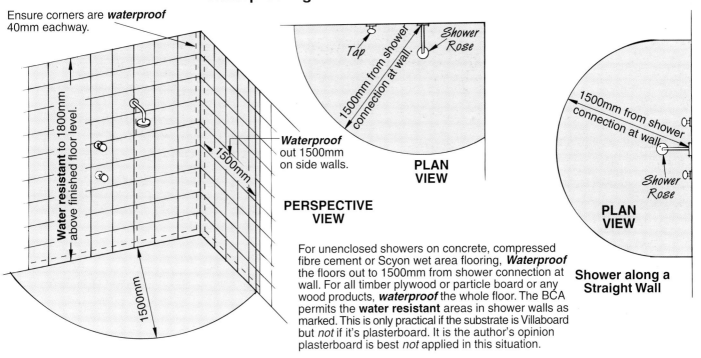

Ensure corners are **waterproof** 40mm eachway.

Water resistant to 1800mm above finished floor level.

1500mm

1500mm

Waterproof out 1500mm on side walls.

Tap

Shower Rose

1500mm from shower connection at wall

PLAN VIEW

PERSPECTIVE VIEW

1500mm from shower connection at wall

Shower Rose

PLAN VIEW

Shower along a Straight Wall

For unenclosed showers on concrete, compressed fibre cement or Scyon wet area flooring, **Waterproof** the floors out to 1500mm from shower connection at wall. For all timber plywood or particle board or any wood products, **waterproof** the whole floor. The BCA permits the **water resistant** areas in shower walls as marked. This is only practical if the substrate is Villaboard but *not* if it's plasterboard. It is the author's opinion plasterboard is best *not* applied in this situation.

Waterproofing of Showers over Baths

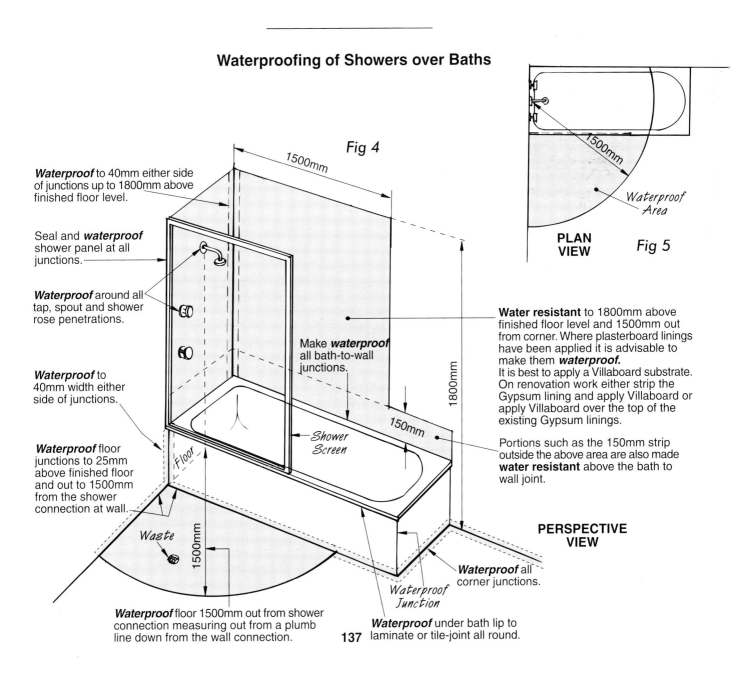

Waterproof to 40mm either side of junctions up to 1800mm above finished floor level.

Seal and **waterproof** shower panel at all junctions.

Waterproof around all tap, spout and shower rose penetrations.

Waterproof to 40mm width either side of junctions.

Waterproof floor junctions to 25mm above finished floor and out to 1500mm from the shower connection at wall.

Waste

Floor

1500mm

Fig 4

1500mm

Make **waterproof** all bath-to-wall junctions.

Shower Screen

150mm

1800mm

Waterproof floor 1500mm out from shower connection measuring out from a plumb line down from the wall connection.

Waterproof Junction

Waterproof under bath lip to laminate or tile-joint all round.

Waterproof all corner junctions.

1500mm

Waterproof Area

PLAN VIEW   **Fig 5**

Water resistant to 1800mm above finished floor level and 1500mm out from corner. Where plasterboard linings have been applied it is advisable to make them **waterproof**.
It is best to apply a Villaboard substrate. On renovation work either strip the Gypsum lining and apply Villaboard or apply Villaboard over the top of the existing Gypsum linings.

Portions such as the 150mm strip outside the above area are also made **water resistant** above the bath to wall joint.

PERSPECTIVE VIEW

137

Preparing a Shower Recess for Tiling

Shower Recesses with a Hob
(see also Pages 136&137)

For showers which have no hob or with set-downs less than 25mm see Pages 136&137.

The recess is waterproofed according to **'Enclosed Showers b. With a Hob' Page 136.** The recess floor is waterproofed and either a silver soldered copper tray, stainless steel tray (water tested), a P.V.C. tray or an approved membrane laid by a certified professional. Copper, stainless steel or plastic trays should be fabricated to fit neatly between all walls. Their sides should be 150mm MIN. high or to 25mm above the maximum possible water retainment level whichever is the greatest see fig 4. Where tiling over or applying sheet laminate to walls, Villaboard® lining should always be applied first.

The tray sides are held in place by the Villaboard® lining (see fig 1). At the doorway the tray extends up to the underside of the tiles and is adhered to the floor and hob with Bostik 'Ultraset' or equivalent. Pre-clean both sides of metal trays with methylated spirits and prime with Bostik 'N40 Primer' or equivalent then apply sufficient adhesive to the underside to hold down all flexing of the tray. Weigh down with weights until adhered. While the hob side of the tray is adhered to the hob, the other sides are *not* attached to the walls but are held in place, by the lower edges of the Villaboard. This is to accommodate movement.

Step 1 The tray is installed. The masonry hob should be laid inside the tray and is fully adhered to the side and bottom as above. *Timber is not permitted for hobs.* Adhere a strip of Villaboard® to the tray side. Both tray and Villaboard® could be adhered and clamped to the hob together. Use adhesive as described above.

Step 2 A bed of coarse river sand mixed with cement to a proportion of four parts sand and one part cement and mixed to a stiff consistency is laid in the tray and shaped to an even grade to slope towards the waste ready for tiling. Tiles are then laid on this surface either bedded into this compo while its still wet or laid in an adhesive bed after the compo has dried (at least 24 hours later).

Using Liquid Membranes (see fig 5, Page 139)

Some of these membranes or the installation method adopted are susceptible to cracking should shrinkage occur in any timber framing behind the membrane. To repair after tiling usually requires the complete reconstruction of the floor and wall tiling, lining and membrane. Ensure the waterproofer uses a long term proven product and methods and provides a guarantee. When liquid membranes are used, work should *not* be carried out above them until they have cured for the specified period. These areas should be barricaded off.

USING STAINLESS STEEL, COPPER or PVC TRAYS

Apply thick bead of sealant between Villaboard and tray to prevent capillary action.

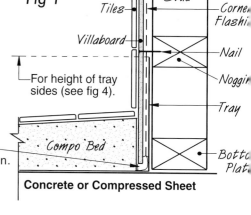

Fig 1

Tiles — Stud — Corne Flashi
Villaboard — Nail
For height of tray sides (see fig 4). — Noggi
Tray
Compo Bed — Botto Plat

Concrete or Compressed Sheet

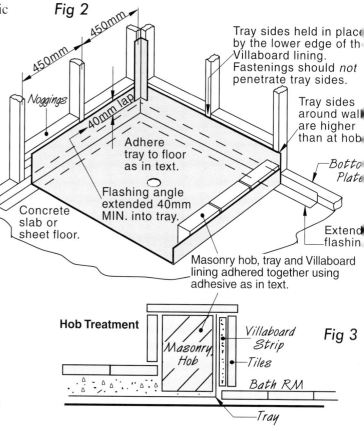

Fig 2

450mm 450mm 450mm
Noggings
40mm lap
Adhere tray to floor as in text.
Flashing angle extended 40mm MIN. into tray.
Concrete slab or sheet floor.

Tray sides held in place by the lower edge of th Villaboard lining. Fastenings should *not* penetrate tray sides.

Tray sides around wall are higher than at hob

Botto Plate

Extend flashin

Masonry hob, tray and Villaboard lining adhered together using adhesive as in text.

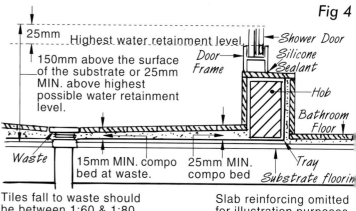

Hob Treatment

Masonry Hob
Villaboard Strip
Tiles
Bath RM
Tray

Fig 3

Fig 4

25mm — Highest water retainment level
150mm above the surface of the substrate or 25mm MIN. above highest possible water retainment level.

Shower Door
Door Frame
Silicone Sealant
Hob
Bathroom Floor
Tray
Substrate floorin

Waste
15mm MIN. compo bed at waste.
25mm MIN. compo bed

Tiles fall to waste should be between 1:60 & 1:80.

Slab reinforcing omitted for illustration purposes.

138

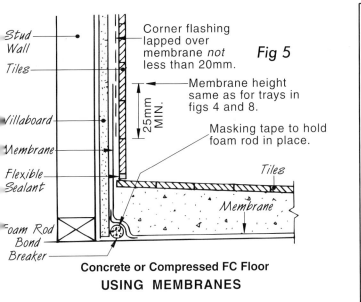

Fig 5

Stud Wall

Tiles

Villaboard

Membrane

Flexible Sealant

Foam Rod Bond Breaker

Corner flashing lapped over membrane *not* less than 20mm.

Membrane height same as for trays in figs 4 and 8.

25mm MIN.

Masking tape to hold foam rod in place.

Tiles

Membrane

Concrete or Compressed FC Floor
USING MEMBRANES

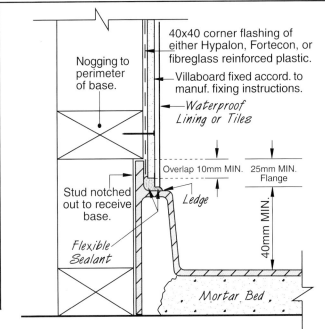

Nogging to perimeter of base.

40x40 corner flashing of either Hypalon, Fortecon, or fibreglass reinforced plastic.

Villaboard fixed accord. to manuf. fixing instructions.

Waterproof Lining or Tiles

Stud notched out to receive base.

Overlap 10mm MIN.

25mm MIN. Flange

Ledge

40mm MIN.

Flexible Sealant

Mortar Bed

Installing Shower Bases (acrylic, polymarble, porcelain moulded steel etc)

Some bases are required to be embedded in compo (a sand and cement mix). Some acrylic bases have acrylic sides and support lugs which enable the base to be simply installed on a level floor.

Installing Waste Pipes. Before commencing, the Plumber should install or check the outlet alignment of the waste pipes.

Step 1 The perimeter sides of the base should be housed into the studs to enable the Villaboard® wall panels to overlap the edges and extend down to within 3-5mm of the base ledge *(see fig 6)*.

Step 2 Measure and mark out the height and thickness of the base sides on the studs. Use a level to transfer the marks around the recess framing.

Step 3 First use a saw to cut the housing depths, except on the corner studs. Then chisel out the housings. Install 75x35 or 50x35mm noggings around the walls between the studs just above the housings.
This is for fastening the lower edges of the Villaboard®.
See also Fig 2, page 114

Step 4 On bases which require bedding into compo, Step 3 is still necessary. The compo is laid and the base is installed (on particleboard or timber floors lay polythene under the wet mortar bed). Check it for level both ways. On bases which *don't* require compo embedment, where necessary, pack up to level with strips of damp proof course or fibre cement. The gap between the lower edges of the Villaboard® lining and the base ledge should be filled with a long life mould resistant sealant such as silicone.

Set-down Shower Recess

For waterproofing details see Pages 136 & 137.
A set-down shower recess is one where the concrete slab has been lowered or stepped down in the shower recess. AS 3740 permits the walls up to 1800mm above finished floor level to be **water resistant**. However it is best to make all shower walls to be *waterproof* including all horizontal and vertical junctions. On step-down recesses as in figs 7 & 8 a copper, stainless steel or PVC tray should be taken up the recess, and across the joint between the floor and framing.
The hob must be masonry and is fully adhered to tray with Bostik Ultraset or equivalent as on Page 138.

Slab reinforcing omitted for illustration purposes.

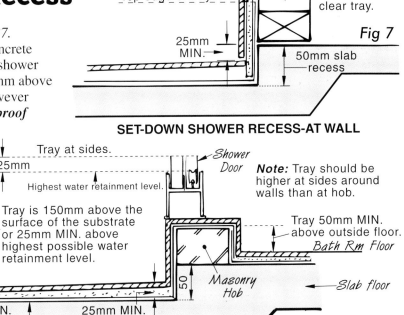

Villaboard

Top edge of tray.

Stud Wall

F.C nail to clear tray.

25mm MIN.

Fig 7

50mm slab recess

SET-DOWN SHOWER RECESS-AT WALL

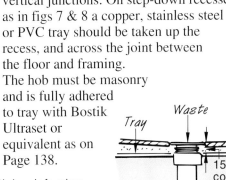

Tray at sides.

25mm

Highest water retainment level.

Tray is 150mm above the surface of the substrate or 25mm MIN. above highest possible water retainment level.

Shower Door

Note: Tray should be higher at sides around walls than at hob.

Tray 50mm MIN. above outside floor.

Bath Rm Floor

Tray

Waste

15mm MIN. compo bed at waste.

25mm MIN. compo bed.

Substrate Flooring

50

Masonry Hob

Slab floor

Fig 8

Tiles fall to waste should be between 1:60 & 1:80.

SET-DOWN SHOWER RECESS-HOB TO WASTE DETAIL

Bath Installation

For baths without Showers
Make **water resistant** around bath to 150mm above bath edge but ensure the joint is *waterproof.*

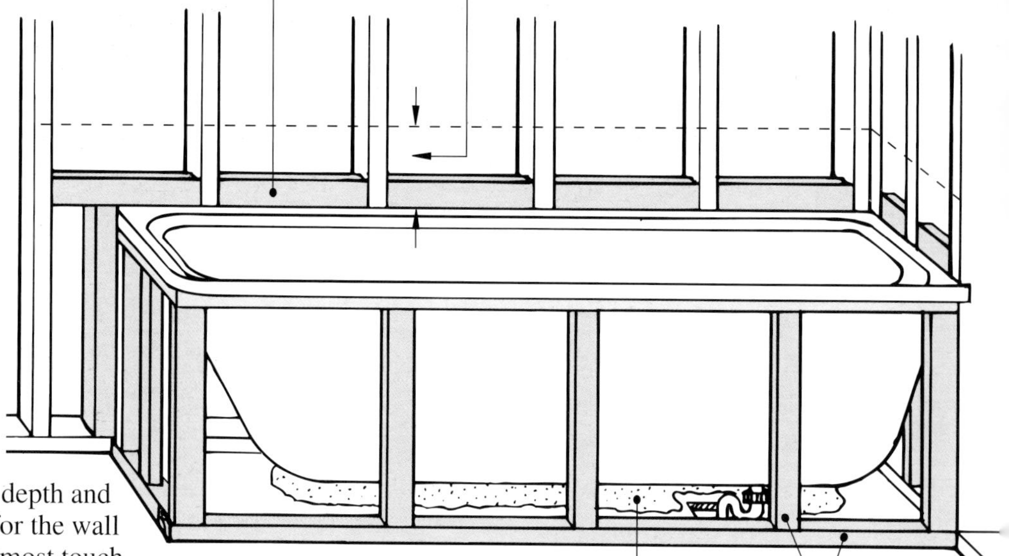

Fitting a Bath

First of all, decide on the height preferred for the top edge. This is usually between 450 and 600mm. If tiles are applied to the side, keep the bath to even tile height to save cutting tiles.

Step 1 Cut a hole 150x150mm in the floor directly below the bath waste for the Plumber to connect waste pipes later (avoid joists). Then on the wall sides of the bath, mark a level line indicating the bath edge (see fig 2). Notch out the studs to the depth and thickness of the bath lip. Allow for the wall sheeting to slide past the lip to almost touch the bath. Then attach a 70x35mm batten on the studs along the level line.

Step 2 Fit noggings between the studs above the notching to provide a fixing base for the bottom edge of wall linings.

Step 3 Slide the bath into the housings and construct the supporting walls using 70x35mm framing on edge. Keep the framework back sufficiently from the bath lip to allow for the lining and tiles to slide behind the lip. Bath sides are usually lined with 6mm Villaboard, then tiled, or laminated panelling is applied.

Fig 1

Noggs around bath edge.

Do not fasten plates through liquid membrane surfaced floors.

Stiff Compo Bedding

70x35mm framing.

Ensure perimeter of bath is **water resistant.** *For waterproof requirements around baths which include a shower over, see the BCA.* Have Plumber connect the waste trap. Some baths have underneath support factory fitted. For those without this support it will be necessary to pack stiff compo (four sand to one cement) underneath along its length.

Allow a gap between the bottom row of tiles and the bath and fill with Dow Corning 'Plumbers Silicone 780' or similar. *Do not use tile grout.*

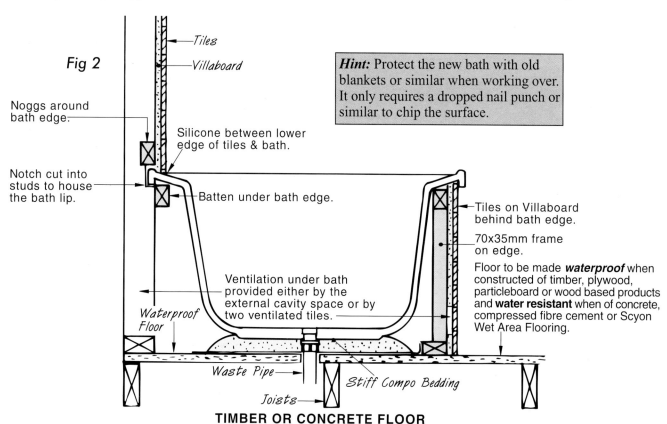

Fig 2

Tiles

Villaboard

Noggs around bath edge.

Silicone between lower edge of tiles & bath.

Notch cut into studs to house the bath lip.

Batten under bath edge.

Tiles on Villaboard behind bath edge.

70x35mm frame on edge.

Floor to be made **waterproof** when constructed of timber, plywood, particleboard or wood based products and **water resistant** when of concrete, compressed fibre cement or Scyon Wet Area Flooring.

Ventilation under bath provided either by the external cavity space or by two ventilated tiles.

Waterproof Floor

Waste Pipe

Stiff Compo Bedding

Joists

TIMBER OR CONCRETE FLOOR

Hint: Protect the new bath with old blankets or similar when working over. It only requires a dropped nail punch or similar to chip the surface.

Miscellaneous

7

Termite Control

All termite control measures should be applied according to AS 3660.1. It is important that the builder understands the need for termite control and the prevention methods available prior to having house plans drafted. This can have a bearing on the construction techniques adopted.

About Termites

Termites can cause extensive damage to house timbers if prevention methods are *not* employed. Termites are 3-7mm long. They build earthen tube-like tunnels to move through. These may lead to holes and cavities in walls such as weep holes or cracks in brickwork. They can operate undetected. For this reason, it is essential that the builder and home owner are confident in the prevention methods adopted and how they are applied. If there are faults or gaps in the applied method, termites are likely to find them. So it is pointless to have a good system of protection if it is *not* properly applied.

Prevention Methods

Houses with Timber Floors

Houses with timber floors should have ant capping (shielding) below the underside of all bearers or plates. *See Page 65 for further details.*

Houses with Slab Floors

Highly toxic chemicals such as organo chlorides have been replaced with low toxic chemicals which must be repeatedly applied to remain effective. This repeatable expense has made long term chemical control an inferior option compared to alternatives.

There are basically four other alternatives:

1. To Build the Structure to be inherently self Termite Resistant — This option requires 75mm of the slab floor perimeter edges to be exposed and unrendered above ground either vertically or horizontally *(see Pages 15 & 16 for details).* Owners should be informed that this concrete edge should always be left exposed and be regularly inspected for termite activity. When this method is applied, it is best to provide paving around the perimeter up to the

> *Note:* The 'BCA' requires a durable notice to be fixed in the meter box, some states may require a second notice fastened inside kitchen cupboards. It should state the **method** of termite protection applied and must also include: The **date** of installation, **where** any chemical barrier was used and its **life expectancy** as listed on the national register of authority label, as well as the manufacturer's **recommendations** for the scope and frequency of future termite inspections and treatments. *Important:* Variations exist from state to state, so check with your local government.

75mm barrier to prevent gardens or bark chip mulch contacting the walls.

Floor Penetrations & Joints — With the above method, service penetrations through the floor such as pipes or other fittings should also be termite proofed. These can be commonly sealed with stainless steel mesh applications.

2. Using Termite Resistant Materials —
The framing can be of treated timbers, cypress, steel, or alternatively the walls can be of masonry and roof framing of either treated timber, cypress or steel. It is considered that while termites may attack mouldings, their replacement *can't* be compared to having to demolish and replace the structural house frame and/or the brick veneer. Ant capping alone on slab floor edges is *not* an acceptable method.

3. Stainless Steel Mesh — Applied by certified applicators and accompanied by a certification or guarantee.

4. Granular Barrier — Applied by certified applicators and accompanied by a certification or guarantee.

Inspection for Termite Activity

Make periodic checks to the outside walls and underneath the house for signs of termite activity. Storing of old timbers and dampness underneath a house are an encouragement for termite activity. At the first sign of termite infestation, call an experienced pest control firm to check and treat.

Termite Traps Laid by Exterminator

These serve as an indicator of termite presence but do not prevent attack.

Arches in Brickwork

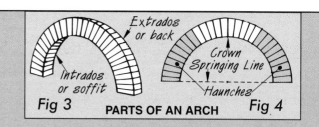

Fig 3   **PARTS OF AN ARCH**   Fig 4

Extrados or back · Intrados or soffit · Crown · Springing Line · Haunches

Arch Centres

The arch formwork supporting a brick arch is known as a 'centre'.

Constructing Semi-circular Arch Centres

Step 1 Cut two arches out of 16 or 18mm particleboard ensuring both are identical (see fig 5). The arch can be marked out with a string attached to a nail and pencil as in fig 1.

Step 2 Attach timber spacer and props of the required depth to space the particleboard out to the exact brickwork thickness (see figs 5 & 7).

Step 3 Nail a strip of thin hardboard over the curved particleboard edges enclosing the space between as in fig 7.

Step 4 Fit the arch centre into the brick opening ensuring it is flush with the brickwork on both sides and that the centreline is plumb. Level the centre by packing up either leg. When all is level and plumb, brace and secure as in fig 7. For brickwork of greater thickness, just increase the depth of the timber props inside the centres with packing.

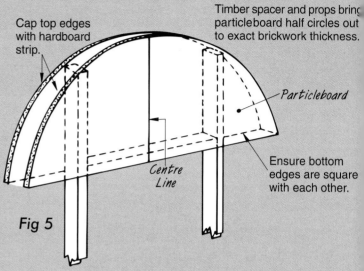

Cap top edges with hardboard strip.

Timber spacer and props bring particleboard half circles out to exact brickwork thickness.

Particleboard

Centre Line

Ensure bottom edges are square with each other.

Fig 5

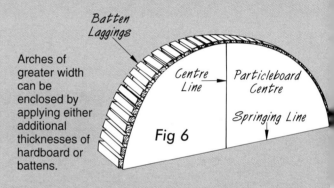

Batten Laggings

Arches of greater width can be enclosed by applying either additional thicknesses of hardboard or battens.

Centre Line · Particleboard Centre · Springing Line

Fig 6

MARKING OUT THE ARCH

Opening Width

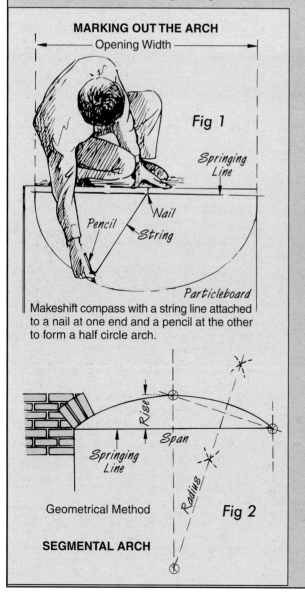

Fig 1

Springing Line

Nail

Pencil

String

Particleboard

Makeshift compass with a string line attached to a nail at one end and a pencil at the other to form a half circle arch.

Rise · Span · Springing Line · Radius

Geometrical Method   **Fig 2**

SEGMENTAL ARCH

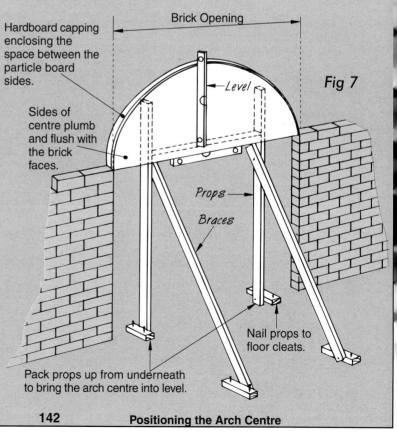

Brick Opening

Hardboard capping enclosing the space between the particle board sides.

Level

Fig 7

Sides of centre plumb and flush with the brick faces.

Props

Braces

Nail props to floor cleats.

Pack props up from underneath to bring the arch centre into level.

Suspended Concrete Patios

Steps to Constructing a Suspended Concrete Patio

Step 1 Have brick columns constructed and their height terminated directly below the underside of the proposed concrete patio.

Step 2 Erect bearers on top of scaffolding jacks or 100x100mm timber toms and brace both ways. Secure support joists and lay sheet formwork. Cut sheets to fit neatly around columns. Nail edge formwork to the flat sheet decking formwork. Fit edge chocks to prevent overturning or bulging while concrete is being poured.

Step 3 Place floor reinforcing mesh, then fit anchor rods into brick columns and bend and tie into mesh. Steel sizes and quantities are obtained from plans or specifications.

Step 4 Pour concrete and remove formwork after fourteen days or according to specifications.

Note: Concrete patio, columns & reinforcing should be designed by an engineer.

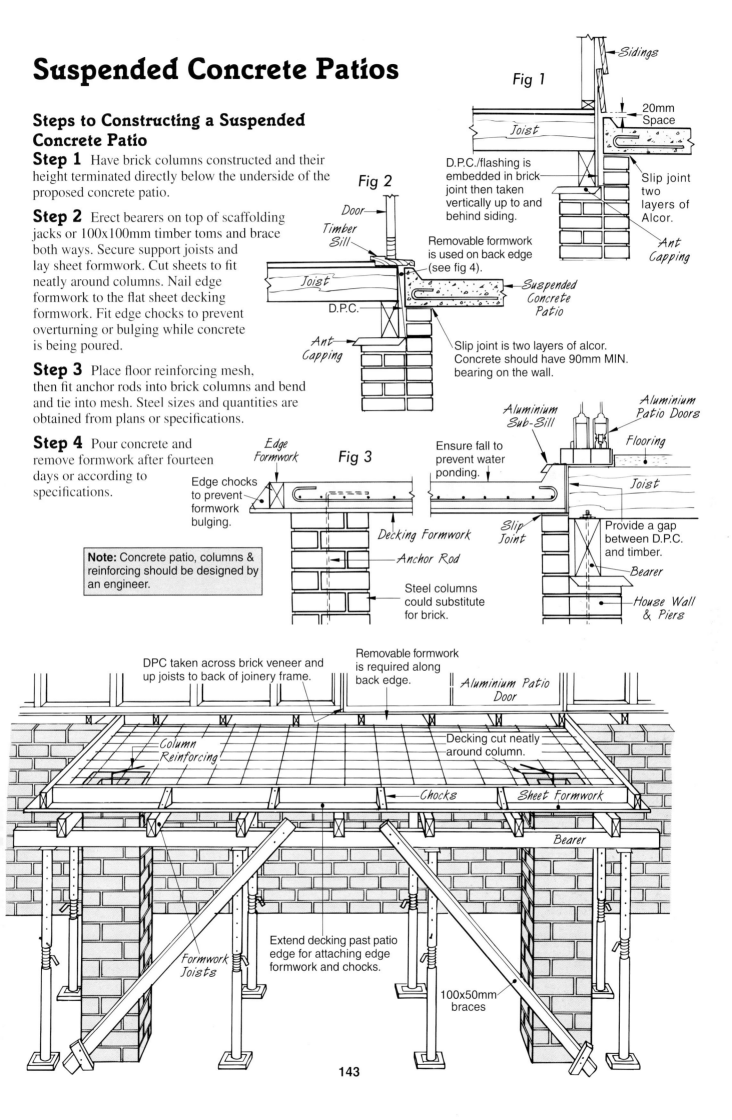

Fig 1

Sidings

20mm Space

Joist

D.P.C./flashing is embedded in brick joint then taken vertically up to and behind siding.

Slip joint two layers of Alcor.

Ant Capping

Fig 2

Door

Timber Sill

Joist

D.P.C.

Ant Capping

Removable formwork is used on back edge (see fig 4).

Suspended Concrete Patio

Slip joint is two layers of alcor. Concrete should have 90mm MIN. bearing on the wall.

Fig 3

Edge Formwork

Edge chocks to prevent formwork bulging.

Decking Formwork

Anchor Rod

Steel columns could substitute for brick.

Aluminium Sub-Sill

Aluminium Patio Doors

Flooring

Ensure fall to prevent water ponding.

Joist

Slip Joint

Provide a gap between D.P.C. and timber.

Bearer

House Wall & Piers

DPC taken across brick veneer and up joists to back of joinery frame.

Removable formwork is required along back edge.

Aluminium Patio Door

Column Reinforcing

Decking cut neatly around column.

Chocks

Sheet Formwork

Bearer

Formwork Joists

Extend decking past patio edge for attaching edge formwork and chocks.

100x50mm braces

143

How to Build & Waterproof Retaining Walls

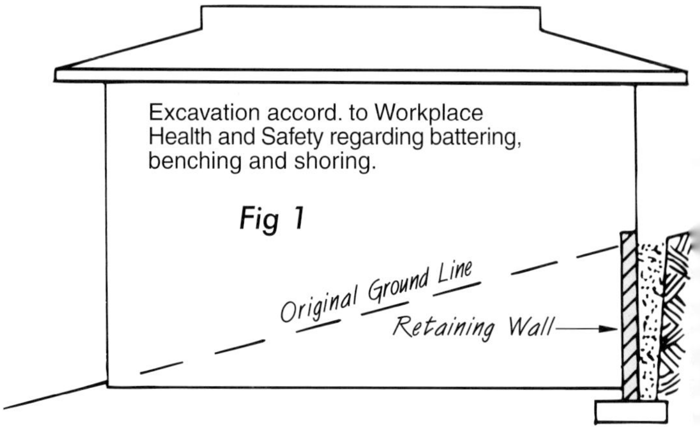

Retaining walls are used where outside earth loads are imposed against the wall. These walls should be designed by an engineer. Where concrete block masonry is used, manufacturers may provide engineer designed plans free of charge. Where one side of a retaining wall is part of a living or basement area, it is important that the waterproofing measures are carefully carried out prior to back-filling to guarantee a permanently dry wall.

Drainage Methods

Two drainage methods are possible: Cordrain and the traditional method (described in full here). Both should be costed out to find which proves the more economical for the situation.

Cordrain & Stripdrain

Cordrain is a plastic drainage in fabric form approximately 20mm thick and comes in rolls 1.2m wide x25m long and can be joined. It is laid up against the wall and allows the excavated material to be backfilled, eliminating the need for gravel, plastic membrane and fibre-cement sheeting. It will save labour costs considerably. **The retaining wall will still require waterproofing as described in Step 3.**

Nylex Polymer Products can supply additional information on Cordrain.

The Traditional Method

The following includes the steps to constructing the wall using concrete masonry blocks.

Step 1 The excavation should be taken beyond the wall sufficiently to allow a person to work in the space between the excavation and the proposed wall in order to waterproof the walls and lay drainage lines.

Step 2 Footings are excavated and reinforcing laid. When concrete masonry blocks are used, vertical reinforcing is tied to align with a string line and positioned carefully to ensure each vertical bar will arrive within the masonry block cores.
Walls are then constructed, reinforced and concrete grout poured.

Step 3 The external wall surface should be waterproofed with two, preferably three coats of Hydroseal or similar. Alternatively use Hydroflex. Hydroseal normally requires one day drying for each coat. Each coat should thoroughly cover without pinholes showing.

Step 4 As a double waterproofing membrane, a layer of polythene is draped over the wall with all joins sealed with pressure sensitive tape. To prevent the gravel fill from puncturing the

polythene, sheets of 4.5mm fibre-cement are laid against the polythene and left permanently in place.

Step 5 An agricultural drainage pipeline should be laid at the base of the wall below floor level. Smooth, slotted agricultural pipe is recognised to be more efficient than the corrugated one.
On retaining walls over 1800mm high, a second pipeline is used about halfway up. These should be carefully bedded and surrounded by 10mm gravel to ensure seepage holes will never become blocked by earth or clay. Ensure a minimum fall of 1:40.

Step 6 Then 10mm gravel is applied against the wall up to ground surface and approximately 200-300mm out from the wall. *Do not top dress.* It is best to lay a dish drain on top.

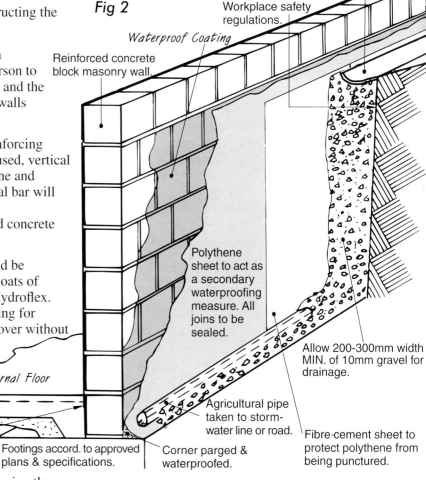

Fig 1

Excavation accord. to Workplace Health and Safety regarding battering, benching and shoring.

Original Ground Line

Retaining Wall

Fig 2

Dish Drain

Excavation accord. to Workplace safety regulations.

Waterproof Coating

Reinforced concrete block masonry wall.

Polythene sheet to act as a secondary waterproofing measure. All joins to be sealed.

Allow 200-300mm width MIN. of 10mm gravel for drainage.

Internal Floor

Moisture barrier returned up slab edge.

Footings accord. to approved plans & specifications.

Corner parged & waterproofed.

Agricultural pipe taken to storm-water line or road.

Fibre-cement sheet to protect polythene from being punctured.

Thin Section Rendered Coatings

Thin section rendered coatings currently is an unregulated industry. Whilst some Manufacturers have training programmes, the trade lacks National Standards. Applicators are *not* always Solid Plasterer tradespeople in fact the majority are not and some current practices are questionable and will ensure problems with coatings will continue to manifest for builders and owners.

Following is a broad overview:

Substrates *The most commonly used are:*
Polystyrene Panel (EPS); **QT Eco Series Wall Panels**; **FC Base Sheets** such as Harditex (Blueboard) or Cemintel Texture Base Sheets; **Hebel Panels and Blocks, Brick and Concrete Block**.

Polystyrene Panel (EPS) Substrate
When choosing this substrate it must be kept in mind that traditionally the exterior walls of a typical dwelling consisted of a structural frame with weatherboard claddings which also provided a bracing value. Its surface was also impact resistant as is brick veneer. In the case of brick veneer a cavity was also present providing a breathing space for the timber frame. On using polystyrene substrates, the timber wall has the layer of polystyrene attached directly over the sarking. In many cases, there is *no* cavity and the polystyrene has no bracing value. Where the coatings have *not* been applied thickly enough hail has proved to cause serious damage to panels.

Installation: Only use expanded polystyrene, not extruded. An impervious vapour barrier must be applied first across the frame because with some situations the polystyrene causes condensation on the rear side especially in air conditioned space. This can result in damage to the frame. However, this will also restrict the frame from breathing. After installation and where exposure to the elements has caused yellowing or surface deterioration, the panels must be broomed off before applying coatings.

Flashings and fastenings: Details are too numerous to include here. *See manufacturer's specifications.*

QT Eco Series Wall Panels (Substrate)
This is a reconstituted polystyrene panel which also contains cement and sand. It is vermin and fireproof. It requires the frame to be battened, permeable sarking attached and weepholes inserted. The system is superior to the above 'EPS' one if manufacturer's recommendations are followed.

FC Base Sheets such as Harditex (Blueboard) or Cemintel Substrate
Used as part of the manufacturer's dedicated coating system. Or applied as a base sheet for acrylic renders only. *Do not use cement/sand based polymer modified renders over this substrate.* Check manufacturer's recommendations regarding coatings. *See also Page 117 for 'Installing Harditex Base Sheet'.*
Areas in which to take particular care:
Vapour permeable barrier This must be provided behind the sheets as recommended.

(Heavy bodied acrylics and Polymer modified cement and sand render).

Fastenings, Control joints and Flashings must be provided as specified. 6mm wide sealant filled gaps must be provided at internal corners. It is advisable to avoid dark colours in finished coatings as heating up of the substrate could result in movement in the base sheet.

Brick and Concrete Block Substrate
These are the most stable substrates. Joints are flush finished and the wall wetted down to create neutral suction for the first coat. Fibreglass, while *not* generally applied, will provide additional protection against future cracking due to movement especially if applied across the entire surface.

Hebel Panels or Blocks Substrate
Refer to the appropriate CSR Hebel instruction booklet.

Fibreglass Mesh for All Substrates
This should be 15oz and have a hole size of 10x10. This mesh is applied across the entire surface of polystyrenes and is a good practice for all substrates. Apply according to manufacturers specifications.

Trim and Corner Mouldings
Stainless steel while expensive, is the preferred trim as plastic has a limited life-span. After renders have been applied, exposed corners or portions of trim must be painted with a full paint system to protect from the elements.

Fastening as specified by the manufacturer.

Coatings (Polymer modified cement/sand or lime) Usually manufacturers make their own formula. There is no specific Australian Standard for them or the consumer to rely upon at present. Mix according to manufacturers requirements. Substrates are generally wetted to prevent undue suction on the first coat. It is usually applied 4-6mm thick. Fibreglass mesh (when applied) is then trowelled into the wet render in vertical runs working from the centre out.
A 2-4mm thick top coat is then applied encasing the mesh then screeded or floated to the specified finish.

Curing the Render 3-4 days is required. If hot, dry or windy conditions prevail, the render is wetted down continually for the first 2 days. The surface then receives its acrylic texture or a quality acrylic paint system.

Acrylic Render Systems

These should be applied by the manufacturer's trained applicators. The coatings are generally applied over Harditex or Cemintel base sheets with fibreglass mesh over joints. However a superior system, though not common, would sheath the entire wall in fibreglass. *This would be performed as follows:* A first coat of acrylic is applied; the mesh then bedded in and a final coat of acrylic applied over the top encasing the mesh. Acrylic coatings are generally *not* thicker than 2mm. They are *not* applied over polystyrene (EPS). They can be applied over previously painted brick or block.

Adhesives

Chemicals have become so sophisticated that it is important to always read the label. To secure manufacturers' warranties use according to their instructions. Follow safety instructions.

PVA (Poly Vinyl Acetate) *e.g. Aquadhere*

This is a water emulsion glue particular for wood joints and to increase the strength of screwed or nailed joints. Spread sufficient on one surface to be absorbed by the adjoining piece. For hungry joints such as end grains you will probably need to apply to both pieces.

Strength of PVA glued joints is increased when clamped while drying. Interior grades are *not* water resistant. For outdoor furniture etc check that the label says **exterior** e.g. *Aquadhere Exterior.*

When using PVAs on wood joists where decorative stains are to follow, always wipe the joint with a damp sponge before it dries otherwise any residue or smears will become obvious blotches that stains will *not* cover.

Poly Urethanes *e.g. Aquadhere Durabond*

This is used for highly **water resistant** bonds. Joints must be clamped as it expands while curing. Allow to fully cure before removing excessive adhesive.

Construction Adhesive *e.g. HB Fuller max bond or Liquid Nails*

These are gap filling adhesives for general purposes *not* for close fitting joinery or furniture. Solvent and water based (fast grab) versions are available. Both will adhere to wood, particleboard, MDF, fibre-cement, glass, metal, tiles and ceramic, fibreglass concrete, painted surfaces, carpet, most plastics (pretest plastic to determine compatibility). *Not for mirrors.* For mirrors, metal and glass use *Liquid Nails Mirror, Metal, Glass or HB Fuller 790 mirror metal & 790 for glass.* For exterior use, use *Liquid Nails Landscape.*

Note: Water based *Liquid Nails Fast Grab* will also adhere to styrene foam. Bonds strengthen faster if at least one surface is porous (e.g. wood or concrete). For faster grab when using either solvent or water based products, use the contact method. Apply to one surface, press to the other joining surface then remove and wait until the surfaces are tacky *(see label for times)* then firmly rejoin. These adhesives should be used in conjunction with nails or screws for highly stressed bonds.

Contact Adhesive *e.g. Gel Grip or Kwik Grip or HB Fuller Rapid Gel Grip*

These adhesives are used for bonding large flat surfaces such as laminate or panels.

Note: Only use a well ventilated area and wear a respirator. The adhesive is applied to both surfaces, allowed to become touch dry then bonded together.

Once the glued surfaces touch, immediate bonding occurs, so be sure the materials are in desired alignment before contact is made. Join the two surfaces from the centre out or from one edge first

to prevent trapping air bubbles. After joining the surfaces can be hammered into tight bonding from the centre out using a block wrapped in cloth under the hammer.

Epoxy Adhesives *e.g. Araldite*

These are a two part adhesive which require mixing before use. They will *not* adhere to polyethylene or polypropylene. There are a number of types available from fast to slow curing to very clear ones.

Mix carefully according to the label. It's a good idea to retain a little to check when it has cured and to be sure it was properly proportioned and mixed.

It is applied across the surface and then the two surfaces joined. They should *not* be moved until the glue has hardened.

Ceramic Tile Adhesive

Seek advise from a tile shop and read the label.

Sealants

There is a mistaken belief that silicone can be used for all sealant applications. *There are mainly four sealants in common use on the building site:* Silicone, Poly Urethanes, Acrylics and Co-polymers. Each has a specific purpose with characteristics unique for that purpose.

Longevity of Sealants

One major cause reducing the longevity of sealants is UV (Ultra Violet) light. Silicone has the highest resistance. Co-polymers the next best followed by Polyurethanes but longevity is *not* the only consideration.

Silicone

Silicone has the longest life expectancy of all sealants. Resistance to the sun and weather versions are available to use with most materials. It is a *inorganic* material while Co-polymers are *organic*. Remember there are **neutral/non-corrosive/non-acid** types which can be used on concrete, galvanizes, zincalume, copper and brass etc. **Acid types** suit glass and non-reactive metals and **mould resistant types** come both in acid and neutral. None of the mould resistant ones should be used in aquariums or for drinking water. For these purposes use *Selleys Glass Silicone* or *Silicone 401.*

Water based silicones can be painted but this may need to be carried out within the first 48 hours or less, check the label. Paintable silicones revert to unpaintable in a short time. Beware of cheap silicones as they may *not* be 100% silicone. Check with the Manufacturer.

Polyurethane Sealants

Usually applied to expansion joints, between joinery, masonry, fibre-cement sheeting etc where structural movement may occur or where adhesives, paint or waterproof membranes are to be applied over the top. Some urethane sealants are also adhesives, performing dual tasks.

MS are silicone modified organic polymers. Are more expensive but perform similar tasks to Polyurethane sealant but without the Isoeyanate content.

Acrylics *e.g. No More Gaps*

These are water based used in place of traditional rigid 'Spackle fillers'. They will tolerate some movement. For greater flexibility ***No More Gaps Weatherboard*** is available but generally Polyurethanes and silicones are preferred in expansion joints.

Co-polymers *e.g. Selleys All Clear or HB Fuller Easy Clear*

This is an advanced solvent based **organic** rubber. It will adhere to most surfaces but *not* to silicones. Co-polymer is a sticky substance and more difficult to apply than other sealants. Use mineral turpentine as a smoothing medium.

One advantage is it can be used in the rain but *won't* adhere to all wet materials such as old wet metals unpainted woods and concrete. It will adhere to most dry surfaces but will attack polystyrene foam and is *not* recommended for permanent water immersion.

Being a clear substance any future mould growth will be visible but it can be over-painted. It can be used on ceramic tiles, sidings, gutters, downpipes, roofs, flashings, sky lights etc. A wipe with acetone will assist adhesion on clean wet surfaces.

How to Apply Sealants

Cut nozzle at 45° and apply with a caulking gun, forcing sealant into the joint to get good contact with the sides. Pushing the nozzle forward with sealant rolling in front of it best achieves this. Immediately smooth (tool) the surface forcing the sealant into the joint with a spatula or other tool recommended on the label. For a neater finish where *not* over-painting use masking tape to prevent over-spread.

Safety on the Job

First Aid Kit A first aid kit must be available on-site and sub-contractors should provide one for their workers.

Fire Extinguishers A fire extinguisher should be easily accessible to extinguish electrical and combustible fires.

Protruding Objects Protruding objects which could cause impaling or abrasion should be covered or bent over.

Asbestos Removal Must always be carried out by licenced persons. A P2 Class respirator should be worn and the asbestos wetted down or encapsulated with an approved encapsulation preparation. The asbestos sheets should be stacked with minimal breakage then wrapped on plastic. Disposal must be at the designated municipal council asbestos disposal site.

Eye, Ear & Breathing Protection Should be provided by the sub-contractor and the principal contractor has an obligation to ensure other trades are *not* affected by the excessive noise of one or more contractors.

Welding on-site When welding on-site appropriate signs must be displayed and eye protection screens erected to shield other workers from welding flash.

Explosive Powered Tools, Nail Guns and Compressed Air Tools Should only be used by trained persons. Warning signs must be erected when in use.

Excavations, Trenches & Confined

Spaces Trenches deeper than 1.5m must be battered, benched or shored and open trenches should have barriers.

Scaffolding

Scaffolding should be constructed according to the 'AS 1576' and in accordance with 'Depts of Workplace Health and Safety'. Scaffolds should be individually designed to suit the specific building and site. Hire establishments usually have trained staff to design and erect. Scaffolding to a height of 4m can be erected by a competent person. Scaffolding in excess of 4m can only be erected by a person with the relevant scaffolding certification. Minimum widths of working platforms (planks) to be 450mm.

Working at Heights

Domestic: Persons working above a clear fall of 3m or more must have either:

1). Properly certified edge protection in place, or

2). Scaffold must extend to afford equivalent protection to certified edge protection, or

3). The use of safety harness, incorporating fall arrest systems, should be worn and properly secured to appropriate anchorages using shock absorbing lanyards.

Domestic & Commercial: On construction work or work on roofs of a slope of over 26° the maximum fall distance allowable is 2m.

Avoiding Ladder Accidents

a). Ladders on-site must have an industrial rating. Ladders should *never* be used to support a working platform.

b). Ladders should slope approx. 300mm away from walls for every 1.2m in height (see fig 4). When working on a roof, the top should extend approx. 1.0m MIN. above the gutter.

c). Step ladders should always have the spacer bar or rope connected.

d). Ladders should be tied at the top and the bottom ends fixed to prevent movement.

Electrical Portable tools should be tested every 3 months and have a current tag. *Do not* use cut, crushed or damaged leads, plugs, sockets or switches. Earth leakage circuit breakers should be installed and emergency shut down equipment in position.

Dogs are *not* allowed on a construction site.

Safety Hints

1. Check the security of all scaffolds, platforms, walkways and ladders before using them.

2. Lock up dangerous objects before leaving the site.

3. Remove protruding nails from timber.

4. Keep electrical cables out of the rain or water. This is a major cause of casualties on building sites.

5. When working with treated timbers follow the safety precautions on Page 8.

Safety While Using Power Tools

a). Wear safety glasses and where necessary, dust masks. On noisy equipment, wear ear muffs. On rotating equipment remove dangling jewellery, tie up long hair and avoid wearing loose clothing.

b). Disconnect the power plug to work on the tool.

c). Remove chuck keys before switching the tool on.

d). Secure the material prior to cutting or drilling.

e). Ensure guards are always operating on power driven cutting equipment.

f). Beware of overconfidence and familiarity when using power driven cutting and grinding equipment.

Avoiding Internal Injuries (lung disease)
Wear a respirator with the appropriate cartridges when working with:

a). Hydrochloric acid, chlorine and other toxic chemicals, paint strippers, polyurethane, epoxy paints and other toxic paints.

b). When sawing or sanding wood, concrete masonry, or fibre-cement sheeting.

c). When working with fibreglass insulation or other

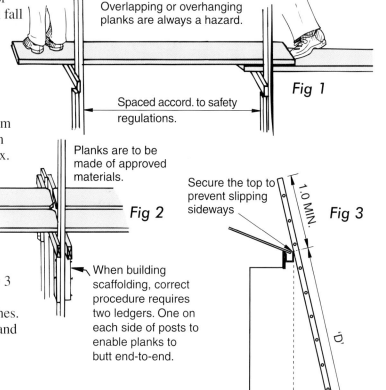

Fig 1 — Overlapping or overhanging planks are always a hazard. Spaced accord. to safety regulations.

Fig 2 — Planks are to be made of approved materials. When building scaffolding, correct procedure requires two ledgers. One on each side of posts to enable planks to butt end-to-end.

Fig 3 — Secure the top to prevent slipping sideways. 1.0 MIN. Secure the bottom to prevent movement. 'D'. ¼ of 'D'.

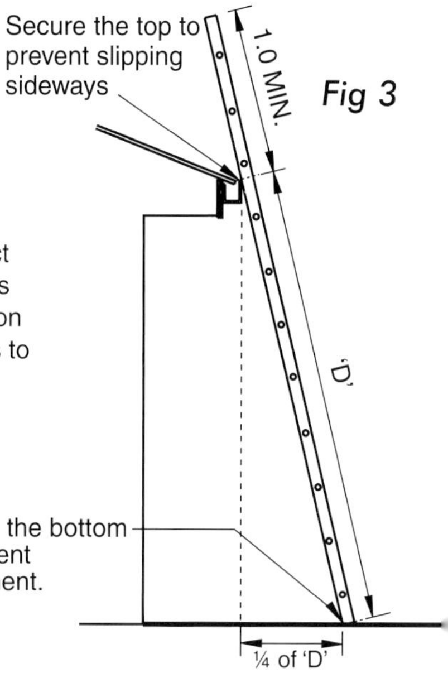

thermal or acoustic insulating materials.

d). When sanding old paint which may have a lead content.

Safely Install Insulation Material
In ceilings provide:

a). Well illuminated working area,

b). Planking over joists to walk on,

c). A permanent means of keeping insulation 200mm clear of all installed electrical fittings,

d). Keep the tops of ceiling joists exposed. Covering joist tops prevents future workman from safe access & movement.

For fibre glass users (from Fletcher Insulation & NSW Goverment fact sheet):
Wear full skin and head covering & use eye protection & a P1 or P2 disposable dust respirator. When in contact with eyes rinse immediately with plenty of water and for itching skin wash with soap & water.

Avoiding Eye Injuries
Wear goggles when:

a). Working with power driven blades or grinders.

b). Working with Hydrochloric and other acids, paint strippers and chemicals.

c). Driving nails and wherever steel hammers impact on masonry or other tools and cold chisels.

d). When welding, remember to wear the correct welding goggles or shield.

Portions of the safety material kindly prepared by Brian Peach of the Peach Institute.

148

Designing for Bushfire Prone Areas *(See also A.S.3959)*

Buildings (designated) bushfire-prone area must be constructed in accordance with the BCA for the assessed level of bushfire threat. There are state variations. Contact your local building approval authority to find out whether your proposed building site is in a declared bushfire-prone area. Appropriate *siting, landscaping, building design, construction methods* and *choice of materials* are all important in reducing the threat and impact of bushfire attack.

The following material is suggested as a guide only.

Construction

The 2009 edition of AS 3959—2009 Construction of buildings in bushfire-prone areas provides two methods of assessing the BAL (Bushfire Attack Level) of a site. There are six levels of risk viz: BAL—LOW, BAL—12.5, BAL—19, BAL—29, BAL—40 and BAL—FZ.

The BAL system (of Bushfire Attack Levels) is based on the potential exposure of the site to heat thresholds, expressed as kW/m2. For example BAL—19 is primarily concerned with protection from ember attack and radiant heat greater than 12.5 kW/m2 up to and including 19 kW/m2.

Note: 19 kW/m2 means 19, 000 watts of radiant heat per square metre.

AS 3959 provides minimum construction requirements for each element of construction for each of the Bushfire Attack Levels (BALs) by providing a range of deemed to satisfy options. The greater the assessed risk, the higher the BAL and the more restricted is the range of deemed to satisfy options. Many of the options provide for both a choice of material AND for a choice of construction or design. Some examples are:

The Floor—For all BALs there is an option of a concrete slab or an elevated floor. If the subfloor space beneath an elevated floor is appropriately enclosed, there are no requirements for the subfloor supports or for the flooring. If it is not appropriately enclosed, there are requirements in the higher BALs.

The Walls—Options include full brick, brick veneer, timber or fibre-cement clad stud walls. For the higher BALs, there are fewer options.

The Roof—Options include non-combustible roof tiles and non-combustible roof sheets. Roofs should be fully sarked. There should be no gaps in the roof covering greater than 3.0mm and it must be fixed properly. Remember, you can have a fireproof house but if a roof tile or sheet is ripped off, it will provide easy access for hot embers.

Screening for Wall vents, Weepholes and Gaps that are not less than 3.0mm —Options include corrosion-resistant steel, bronze and aluminium mesh with a maximum aperture of 2.0mm. For the higher BALs, Aluminium mesh is not acceptable.

Design

Windows—Minimise the size and number of windows facing the direction from which a bush fire is likely to approach.

Windows and Doors—The options for window and door design and window framing and door material are greater if complying bushfire shutters are installed to protect the window assemblies and doors from ember attack, flying debris, radiant heat and strong winds. Appropriate shutter designs are provided in AS 3959. A complying bushfire shutter is a shutter which, when closed, protects a window or a door and does not have any gaps greater than 3 mm in the shutter or between the shutter and the wall. The materials from which the shutter can be made are specified and are increasingly restricted as the assessed BAL increases. These shutters also provide extra home security.

Roofs—Complicated roof designs provide areas for debris to accumulate and require constant cleaning. Roofs should be designed to reduce or eliminate areas where burning debris can collect during a bushfire event. Wherever possible, avoid valleys in roof design. Hip roofs tend to shed the debris better.

Roof penetrations—Roof penetrations, such as chimneys, roof lights, and vent pipes, should be avoided if possible.

Eaves ventilators—Eaves ventilators should be non-combustible with holes or slots screened with a wire mesh with a maximum aperture of 2.0mm. For the higher BALs, there are fewer options for the type of wire for the mesh.

Roof gutters—Building design should allow easy access to keep gutters free of leaf matter. Use non-combustible leaf guards.

Building shape—Complicated building shapes tend to increase wind turbulence and provide traps for burning particles to accumulate against the building. A square or rectangular shape avoids "internal" corners such as in an L-shaped building. Avoid deep verandahs.

Subfloor spaces—These should be inclosed.

Water Tanks, Ground Covers & Fences

It is highly recommended in bush fire prone areas to have a water tank. This should *not* be plastic or made of a combustible material. Wood, bark chips and plastic ground covers should be avoided. Non combustible fences can be useful as they can form a trap for embers.

For further information see—Standards Australia Handbook HB 330—2009 'Living in Bushfire-Prone Areas' (A guide to reducing the threat and impact of bushfire attack and an explanation of the basis of AS 3959)

Energy Efficiency for Dwellings

LOCATION	CLIMATE ZONE
Darwin	1
Brisbane	2
Alice Springs	3
Albury, Moree, Port Augusta	4
Sydney East & Adelaide	5
Perth	5
Sydney West & Melbourne	6
Canberra & Hobart	7
Thredbo	8

The BCA has produced a criteria for establishing the efficient use of energy in buildings and thereby reduce greenhouse gases. These criteria vary according to the geographic location. Energy conserving measures required for a cooler climate will be different from temperate and tropical ones. The BCA divides Australia into eight climate zones, *see chart*. For a detailed chart see the BCA.

Where house plans are submitted to local government for approval; the design, orientation of the house, materials chosen for the structure, plus additional measures such as thermal insulation blankets, window type, shading, building sealing, ventilation etc will provide an energy efficiency value number for the floor, walls, roof and glazing. These values must meet the criteria set down for the particular climate zone. One way of establishing the total energy efficiency is to employ an assessor who uses software. The other is to comply with previsions in the BCA.

The BCA, or Basix in NSW, provides the thermal resistance levels (R values) for each climate zone and suggests building materials and practices for achieving those levels. The following will provide some further information related to achieving the required 'R' value/ 'U' value for each element.

Ground Floors
Zones 1,2&3 have no requirements. Suspended floors and slab-on-ground floors in cooler temperate zones must include heat loss measures i.e. insulation underfloor or slab edge insulation.

External Walls
These are to be of a required mass (thickness, density) and be either shaded for higher temperatures or insulated from hot or cold to achieve the thermal levels required. Various options to achieve the required thermal levels are outlined in the BCA or Basix in NSW.

Roofs
These have upwards and downwards heat flows which have to be insulated against. Zones 1,2&3 have concessions available.

Roof Colour can greatly reduce heat transmission. *Typical absorptance values are as follows;* the lighter the colour the less heat generated: dark grey 0.90, green and red 0.75, buff or yellow 0.60, zinc aluminium-dull 0.55, light grey 0.45, off white 0.35 and light cream 0.30.

Roof Lights
Depending on the location in the dwelling these need to be energy efficient. They can gain excess heat and lose an excess of warmth after sundown. When several are planned, their total area is restricted to reduce heat transmittance.

External Glazing
The following factors determine the energy loss and gains: climate zone, the glazing type, the aspect in relation to sun and shading. Shading may be achieved by - the eaves, blinds or a balcony/verandah above.

Shading
Shading can be achieved by eaves, blinds or verandahs/ balconies. Below latitude 20° (Bowen Qld) shading is *not* required on the south side of dwellings.

Building Sealing
This requirement aims to reduce the amount of air escaping from an air conditioned space in order to reduce the amount of energy required to cool or heat that space. It involves sealing off all gaps or openings by: fitting weather strips around doors and windows, fitting self closing dampers on exhaust fans, dampers to chimneys and floors and sealing gaps around roof lights.

Air Movement
Ventilation openings in external walls are to be a minimum percentage of a room area. Restrictions are stipulated as to the location of openings and how they interconnect with other openings through the dwelling. Hotter climate zones can utilise ceiling fans. Though higher ceiling heights are recommended to avoid accidents.

Domestic Services
Such as air conditioning and central heating duct work is to be sealed to prevent air loss and to also be insulated with the MIN. 'R' value requirement. Central heating water pipelines outside of conditioned spaces must also be insulated.

Installation of Insulating Materials
(reflective fabrics & bulk insulation) Unless these are installed as follows their maximum benefits will be reduced. **Reflective Fabrics:** Overlap joins *not* less than 150mm or tape together. Fit tightly around openings and penetrations; **Bulk Insulation** such as fibreglass wool batts or blankets: must *not* be compressed except where it passes over battens, electrical cables or water pipes or similar elements. Units of insulation must fit tightly up against those adjoining, *see Page 148 for safety precautions*.

Designing & Building for Energy Efficiency

The energy efficient house should be firstly designed & oriented on the site for that purpose. The selection of the site is also a major contributor. When these factors are considered additional elements such as insulation, heating & cooling systems can only enhance the overall efficiency. The first step is the selection of the site. This will, in the main be governed by the second step: The Design of the House. So we'll look at the latter first.

The Design of the House

(In Cool Tempreture Climates)
To maximise free heating of the house by the sun in the winter demands a floor plan layout that has all living areas & if possible, a master bedroom all facing the northern aspect or the passage of the winter sun. As in fig 1. In this perfect situation the plan has a long wall facing north. As house sites may prevent this, solutions could include: Two storey or stepping the design

Other Elements to Consider

Windows (see Fig 1)

Northern Aspect Apply large glass areas along this aspect.

Eastern Aspect Apply glass as necessary, as, while early morning sun may strike this wall it will be overhead before it reaches higher temperatures.

Southern Aspect Keep glass areas to a minimum as they will only cause a loss of heat in winter.

Western Aspect Avoid glass on this aspect if at all possible and never place kitchens on this wall as they generate enough heat without adding solar heat.

Shading

Eaves should be applied on all sides to provide shade & weather protection on walls and windows in summer. 500mm Eave depth should be a minimum and 600mm a maximum. This will prevent most direct sun penetration in summer and emit maximum direct sun rays in winter (see Fig 2.)

Ventilation

Windows & openings in the floor plan layout should enable cool summer breezes to pass from one side of the building to the other. Should any rooms be isolated from these breezes have openings positioned high in walls or above doors to capture them. Louvres or similar could be used-electronically operated or manually with a long handle.

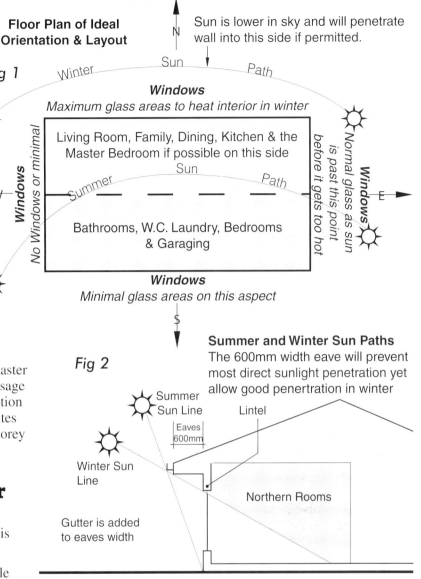

Floor Plan of Ideal Orientation & Layout

Fig 1

Sun is lower in sky and will penetrate wall into this side if permitted.

Winter Sun Path

Windows
Maximum glass areas to heat interior in winter

Living Room, Family, Dining, Kitchen & the Master Bedroom if possible on this side

Summer Sun Path

Bathrooms, W.C. Laundry, Bedrooms & Garaging

Windows
No Windows or minimal

Windows
Normal glass as sun is past this point before it gets too hot

Windows
Minimal glass areas on this aspect

Fig 2

Summer and Winter Sun Paths
The 600mm width eave will prevent most direct sunlight penetration yet allow good penertration in winter

Summer Sun Line

Lintel

Eaves 600mm

Winter Sun Line

Northern Rooms

Gutter is added to eaves width

Ventilators

Hot air rises & accumulates beneath the top of roofs, builds up, then passes back down through ceilings. Wind driven roof ventilators should be placed high in roofs to remove warm air.

Site Selection

Sites facing north and north east are preferred as they receive early morning sun if the rising sun is not interrupted by large buildings or a hill between. But if the site is not on a good aspect a good designer can still make maximum use of solar energy and local breezes by correct orientation & placement of rooms.

Cooling For Dwellings in Climates Not Requiring Heating

a). Orient the house & windows to receive local breezes in living areas,
b). Install windows such as casements or louvers which have large opening ability,
c). Elevate the house to capture breezes,
d). Construct light weight walls which don't retain heat,
e). Install higher ceilings with fans.

151

Estimating Quantities

Some manufacturers and suppliers of the following items often provide a measuring and costing service: Timber and roof trusses, windows and wallboard, roof cladding, bricks, external claddings, kitchen cupboards, plumbing and hardware. Subcontractors will often supply remaining quantities or quotations. However, knowing how to quantify the following items may prove invaluable.

Estimating Concrete Quantities

To find the cubic metres required, multiply the width by the thickness by the length.

A slab footing .5m x .35m x 50m = 8.75 cu.m. Add to this any piers. Allow an extra 7.5% for waste.

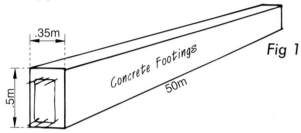

A slab floor 100mm thick x 7.8m x 15m = 11.7 cu.m. (cubic metres). Add to this floor and edge thickenings and edge beams. Allow an extra 5% for waste.

Concrete Block Masonry

Block Quantities: for 200 series blocks (390x190x190): 12½ blocks for every square metre.

Mortar Quantities: 1 cubic metre will lay approx. 800-900 blocks. When using a mix of 1:1:6, a cubic metre will require 6 bags cement, 6 bags lime and 1.2 cubic metre (damp) sand (approx.).

Grout for block masonry walls: Block type 2001 will require 1 cubic metre to fill approx. 150 blocks.

Timber Framing (see also Pages 6-10)
Studs

Indicate with a mark on the floor plan the position of each stud at their correct centres using a scale (see fig 3). Allow three studs to every external and internal corner for weatherboards and two for brick veneer plus studs at sides of windows and secondary jamb studs where required. Designate jack studs above and below windows to be full length common studs. This extra allowance becomes a safety margin. Studs are bought in approximate stud lengths or can be cut to your own requirements.

Plates & Noggings

Using a scale on the floor plan, measure the length of each wall including windows and doors into a total sum. Then multiply this figure by three. If there are double top plates of same size, multiply by four.

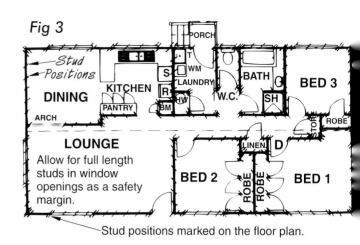

Stud positions marked on the floor plan.

This represents top and bottom plates and one row of noggings. Add a further 10% for wastage.

For example:
Total Wall Lengths — 70m x 3 = 210 + 10% (safety margin for wastage 21 met.) = 231 metres total.

If plates and noggings are of different thicknesses, calculate their lengths separately.

Joists, Bearers & Battens

Again using a scale, mark the position of each member on the floor plan at the spacings stipulated. Include any joists for wall support and any double joists for end walls. Total the quantity. Joists and bearers are bought in their correct length plus wastage. Battens are bought by the lineal metre.

Rafters & Roof Timbers

Rafter lengths will be found in 'The Australian Roof Building Manual' or they can be scaled from a plan, elevation or section and drawn to large scale. The remaining roof timber can also be scaled from the plans. (Always include a little extra for wastage).

Bricks

There are fifty-six standard metric bricks 230x110x76mm to each square metre. This figure includes wastage. Bricks are quantified by multiplying the length of each wall by its height then multiplying the total by fifty-six (see fig 4). For cavity brick, double the figure.

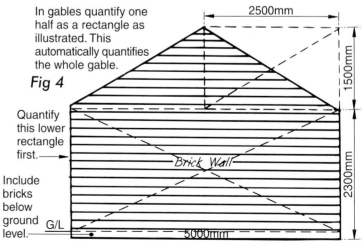

Example: Main Wall 5000x2400mm = 12.00 sq. metres;
Gable 2500x1500mm = 3.75 sq. metres
15.75 x 56 = 882 bricks

Details Plans Should Contain

The following items should be included in plans to prevent on-site disputes and to legally protect the contractor and the owner.

Floor Plans (to 1:100 Scale)

a). Dimensions of overall brickwork, stud framing and room sizes to rough stud frames. Trim openings of all windows and doors. Space allowances for refrigerator and white goods. Wardrobe depths. Location and spacing of all columns and verandah posts.

b). Roof and eave lines as dashed lines.

c). Doors and windows to have a legend reference describing the details of each.

Foundation & Slab Floor Plan

The engineer's design will include: details and specifications regarding footings, slab and thickenings, reinforcement, stirrups and their spacings and concrete strength. These are *not* required on the other drawings.

a). Dashed lines indicating footing excavations and
column, post or verandah post stirrup locations.

b). Location of all plumbing units requiring drainage waste pipe installation. Location of any shower recess to enable exact location of formwork.

c). Brick base wall thickness plus size and spacings of engaged and isolated piers.

d). Concrete slab vapour barrier thickness and reinforcement type and placement. Surface finish requirement of slab and patio floors.

e). Any retaining walls must be fully detailed.

f). Termite protection method.

Bracing Plan

a). The design wind speed and load force required and the values provided in each direction.

b). All bracing clearly lined with a legend if necessary referring to each bracing type. Standard bracing details and standard connection details for each tie down.

Cross Sections

a). These views should be taken through the highest and widest points of the dwelling and should reveal details or facts which are otherwise concealed. The more this drawing reveals the less disputes are encountered with subcontractors and owners. Where necessary, additional sections may be required should the structure greatly differ in other cross sections.

b). Ceiling and door heights can be applied here as well as wall lining types.

Wall Details

a). These are best taken through a typical opening. This should be in 1:20 or larger scale and with as much information as possible.

b). Fastenings and spacings for roof to wall plate, wall-to-floor and floor to footing connections.

c). Brick veneer tie type and their spacings.

d). Placement of flashings, D.P.C.'s, weepholes and spacings and termination point for underslab PVC membrane.

e). Termite protection methods.

Elevations (all elevations should be drawn)

a). Positions of all windows and doors.

b). Heights of ceilings, doors and windows above the floor surface can be marked here.

c). Dashed lines indicating external wall bracing optional on elevations but good practice.

d). Dashed lines indicating natural ground lines (this will enable more accurate calculating of materials below floor level).

e). Roof and wall claddings and finish.

f). Types of glass selected for specific windows and doors.

g). Roof vents, air conditioning units, and solar H.W.S. locations.

Site Plans (to 1:200 Scale)

a). Minimum distances from boundaries.

b). Real property description and lot or street numbers.

c). Roofing downpipe outlets shown as 'DP' and method of disposal.

d). North direction indicated.

e). Street position and owner's name.

f). Contours, bench levels, finished floor level, extent of earthwork, embackments, retaining walls, driveways, site drainage, council infrastructure such as sewer mains and easement.

Using & Reading Plans

a). Attach a set of plans to a panel of ply or hardboard and keep it out of the sun on-site during construction. Consider plastic lamination.

b). When alterations are made during construction be sure to mark these clearly on the plans.

c). Always take measurements from the marked dimensions *not* by scaling off the plan.

d). Room dimensions are always taken from the rough stud edge.

e). Broken lines on internal walls indicate bracing walls.

f). Broken lines on the foundation plan represent footing width.

g). Ground lines are indicated on elevations to reveal the amount of subfloor materials required and as a footing construction guide. If a construction or fastening method is *not* marked on the plans or specifications, check with manufacturer's instructions or with a Local Building Authority.

Typical House Plan Example
(photo reduced)

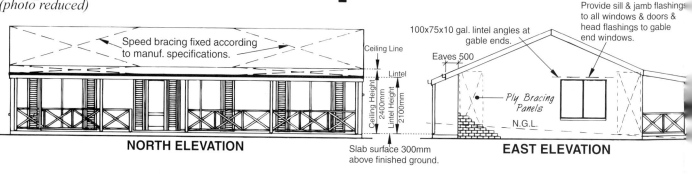

Speed bracing fixed according to manuf. specifications.

Ceiling Line
Lintel
Ceiling Height 2400mm
Lintel Height 2100mm

NORTH ELEVATION

100x75x10 gal. lintel angles at gable ends.

Provide sill & jamb flashings to all windows & doors & head flashings to gable end windows.

Eaves 500

Ply Bracing Panels

N.G.L.

EAST ELEVATION

Slab surface 300mm above finished ground.

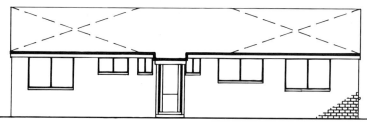

SOUTH ELEVATION

Brick selection by owner.

WEST ELEVATION

WINDOW & DOOR SCHEDULE			DOOR SCHEDULE		
NO.	SIZE	TYPE	DOORS	SIZE	TYPE
W1	1500x1810	Bronze Anodised Sliding Windows	A	2040x870	Sel. Exterior Solid
W2	1200x1810		B	2040x820	Sel. Exterior Hollow Core
W3	.900x1810	" "	C	2040x820	Sel. Sliding Hollow Core
W4	.600x.610	" "	D	2040x770	Sel. Hinged Hollow Core
W5	.600x.610	" "	E	2040x720	" "
W6	.600x1210	" "	F	2040x620	" "
W7	1200x1810	" "			
W8	.600x1810	" "			

ALUMINIUM SLIDING DOORS			W.C. door hinges to be detachable from the outside or to be hinged to open out.
W9	2100X1800	Bronze Anodised Sliding Doors	
W10	2100X1800		
W11	2100X1800	" "	

Verandah rafters and roof trusses @ 900¢ fixed with 1/triple grip at each end with 10/35x3.15mm Pryda nails. 4 to each side leg and 2 to the top plate.

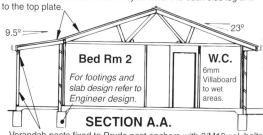

9.5°   23°

Bed Rm 2
For footings and slab design refer to Engineer design.

W.C.
6mm Villaboard to wet areas.

SECTION A.A.

Verandah posts fixed to Pryda post anchors with 2/M12 gal. bolts top and bottom.

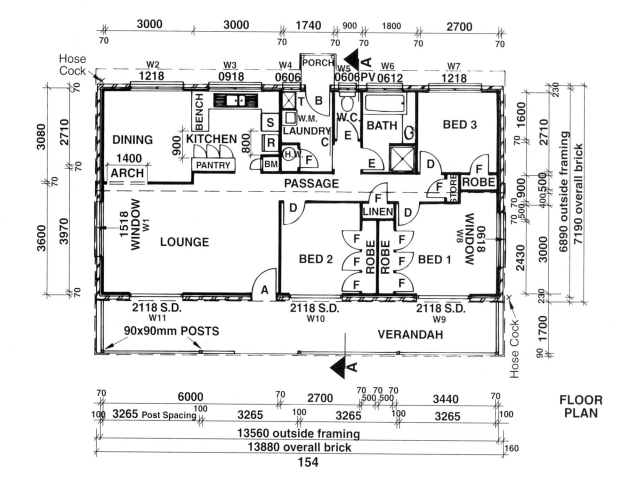

Hose Cock

W2 1218   W3 0918   W4 0606   PORCH   W5 0606 PV   W6 0612   W7 1218

DINING   1400   ARCH

BENCH   KITCHEN   PANTRY   900   800

T   B
S   W.M.
R   H.W.
BM   LAUNDRY
C
E
F

W.C.   BATH   BED 3

LOUNGE   1518 WINDOW W1

PASSAGE

D   LINEN   D
F   ROBE   ROBE   F
BED 2   F   F   BED 1

STORE   D   F   ROBE
0618 WINDOW W8

A

2118 S.D. W11   2118 S.D. W10   2118 S.D. W9

90x90mm POSTS

VERANDAH

Hose Cock

FLOOR PLAN

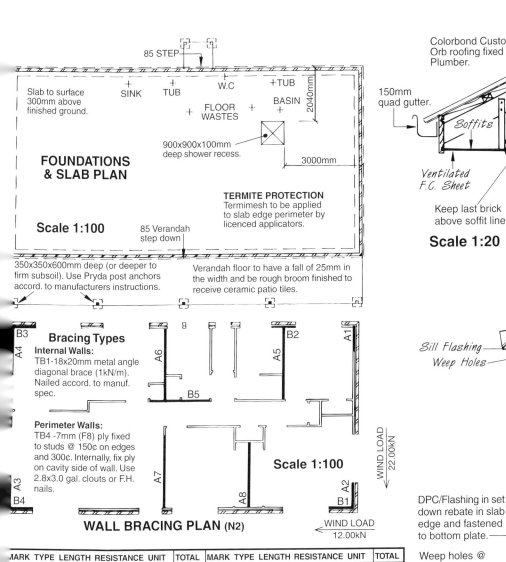

FOUNDATIONS & SLAB PLAN

Scale 1:100

85 STEP

85 Verandah step down

SINK TUB W.C TUB
FLOOR WASTES BASIN

2040mm

Slab to surface 300mm above finished ground.

900x900x100mm deep shower recess.

3000mm

TERMITE PROTECTION
Termimesh to be applied to slab edge perimeter by licenced applicators.

350x350x600mm deep (or deeper to firm subsoil). Use Pryda post anchors accord. to manufacturers instructions.

Verandah floor to have a fall of 25mm in the width and be rough broom finished to receive ceramic patio tiles.

Bracing Types

Internal Walls:
TB1-18x20mm metal angle diagonal brace (1kN/m). Nailed accord. to manuf. spec.

Perimeter Walls:
TB4 -7mm (F8) ply fixed to studs @ 150¢ on edges and 300¢. Internally, fix ply on cavity side of wall. Use 2.8x3.0 gal. clouts or F.H. nails.

B3 A4 B2 A1 A6 A5 B5 A7 A8 A2 B1 A3 B4

Scale 1:100

WIND LOAD 22.00kN

WALL BRACING PLAN (N2)

WIND LOAD 12.00kN

MARK	TYPE	LENGTH	RESISTANCE	UNIT	TOTAL	MARK	TYPE	LENGTH	RESISTANCE	UNIT	TOTAL
A1	TB4	900	2.02	900	2.020	B1	TB4	900	2.02	900	2.020
A2	TB4	900	2.02	900	2.020	B2	TB4	900	2.02	900	2.020
A3	TB4	900	2.02	900	2.020	B3	TB4	900	2.02	900	2.020
A4	TB4	900	2.02	900	2.020	B4	TB4	900	2.02	900	2.020
A5	TB1	2700	2.70	2700	2.700	B5	TB1	2400	2.40	2400	2.400
A6	TB1	2700	2.70	2700	2.700						
A7	TB1	2700	2.70	2700	2.700						
A8	TB1	2400	2.40	2400	2.400						
Nominal Bracing [External Walls Sheeted 1Side-0.3kN/m]					1.860	Nominal Bracing [External Walls Sheeted 1Side-0.3kN/m]					1.950
Nominal Bracing [Internal Walls Sheeted 2Sides-0.5kN/m]					3.300	Nominal Bracing [Internal Walls Sheeted 2Sides-0.5kN/m]					6.700
			Total Resistance		23.740				Total Resistance		19.130
			Total Wind Load		22.000				Total Wind Load		12.000

Colorbond Custom Orb roofing fixed by Plumber.

150mm quad gutter.

Soffits

Ventilated F.C. Sheet

Keep last brick above soffit line.

Scale 1:20

WALL SECTION

Plasterboard Ceiling

75x38 (F11) binders @ third spans.

Sill
Wall Lining

Sill Flashing
Weep Holes

110mm brick veneer fixed to timber frame with med. duty gal. ties @ 610mm crs. vert. and 410 horizontally.

Wall framing, roof and ceiling battens accord. to schedule.

DPC/Flashing in set down rebate in slab edge and fastened to bottom plate.

M10 anchor bolts to bottom plate @ 1200mm ¢ MAX.

F72 reinforcing fabric

Weep holes @ 750¢. MAX.

Paving

Slab Floor

Cavity below ground level filled with grout up to set down.

Sand Bed
200UM PVC membrane

Footing and slab accord. to Engineers drawing. All concrete 20MPa.

TIMBER SPECIFICATION

External Framing [Seasoned F5 Softwood or F11 Hardwood]

Studs @ 450mm ¢ [Notched]   - 70x45 [F5] or 70x35 [F11]
[Not Notched] - 70x45 [F5 or F11]

OPENINGS	STUDS	LINTELS [F5] or	LINTELS [F11]
900	1 common stud	45x70mm	45x70mm
1200	2 common studs	140x35mm	90x35mm
1800	2 common studs	190x35mm	140x35mm

Top Plate [Not Trenched] - 70x70 [F5] or 45x70 [F11]
Bottom Plate [Not Trenched] - 70x70 [F5] or 45x70 [F11]

Internal Framing [Seasoned F5]

Studs @ 450mm ¢ - Top and Bottom plates - 70x35mm

Roof Trusses [to Manu. Spec.] @ 600 or 900mm ¢ fixed as on 'A-A'

Roof Battens [F11]
38x75mm @ 1200 ¢ MAX.   Fixings to each crossing.
[Within 1200mm of Roof Perimeter]
HWD Trusses @ 600mm ¢ - 1/75x3.75 dia. grooved nail.
Pine Trusses @ 600mm ¢ - 2/75x3.75mm dia. grooved nails.
HWD Trusses @ 900mm ¢ - 2/75x3.75mm dia. grooved nails.
Pine Trusses @ 900mm ¢ - 1/75 No.14 Type 17 screw.
[General]
Pine Trusses @ 900mm ¢ - 2/75x3.75mm dia. grooved nails.
[Remainder] @ 900mm ¢ - 1/75x3.75mm dia. grooved nail.

FLOOR AREAS

House	100.07	sq. m.
Verandah	24.98	sq. m.
Porch	1.14	sq. m.
Total	126.19	sq. m.

Real Property Description:
Lot No. :

42.00

1500
D.P.

Contour Lines

Wall Line

N

Eave Line

Proposed

Dwelling

30000mm

Side Boundary

SITE PLAN

42.50

Sewer Line

10000mm

Eave Line

Driveway

D.P.

25000mm

43.00

Front Boundary

Scale 1:200

Proposed Residence for.....................
At ....

Scales 1:100 (except where otherwise indicated).
Wind Load   N2
Measurements take precedence over scale. Check measurements on-site.
All construction work to be in accordance with the B.C.A. and the specific Australian Standards and Local Council requirements.

Common Abbreviations & Symbols

@	at	ORG	Overflow Relief Gulleys
accord.	according	PERP	Perpends (vertical block or brick mortar joint)
Approx.	Approximately		
AS	Australian Standard Codes	PVC	Polyvinyl Chloride
BCA	Building Code of Australia	PERP	Perpends or vertical block or brick joint
BK	Brick		
BWK	Brickwork	QUAD	Quadrant Moulding
¢	centre to centre	REINF.	Reinforcing
C1,C2,C3,C4	Cyclonic Wind Categories	RH Bolts	Round Head Bolts
CCA	Copper Chrome Arsenate	RHS	Round Hollow Section (steel)
c/s	countersunk	RPD	Real Property Description
Dia.	Diameter	SD	Sliding Door
DPC	Damp Proof Course	SHS	Square Hollow Section (steel)
DW	Dishwasher	SPEC	Specification
EML	Expanded Metal Lath	Sq.	Square
EJ	Expansion Joint	S.S.	Stainless Steel
FC	Fibre-Cement	SWD	Softwood
FH	Flat Head Nails	T & G	Tongue & Groove
g	gauge (bolts/screws)	W/B or	
GAL	Galvanising	W/Boards	Weatherboards
G/L	Ground Level	WC	Water Closet
H1, H2, H3, H4, H5	Levels of preservative treatment in timber. 'H' = abbreviation for 'Hazard'		
Hex.	Hexagonal head (bolt)		
HSOB	How to be a Successful Owner Builder Renovator manual		
HT	Height		
HWD	Hardwood		
HWS	Hot Water System		
HWR	Shower		
kN	kilo Newtons (force)		
kPa	kilo Pascal (pressure)		
LVL	Laminated Veneered Lumber		
m	Metre		
mm	Millimetres		
M/sec	Metres per second		
MANUF	Manufacturer		
MAX.	Maximum		
MIN.	Minimum		
MPa	Mega Pascal (pressure)		
MS	Mild Steel		
N	North		
N1,N2,N3,N4 N5,N6	Wind Categories		
NGL	Natural Ground Level		
OG	Obscure Glass		

Drainage Abbreviations

BV	Back Vent
CIP	Cast Iron Pipe
EV	Educt Vent
EW	Earthenware
GEW	Glazed Earthenware
GIT	Grease Inspection Trap
GT	Gully Trap
FW	Floor Waste in Drainage
IC	Inspection Chamber
IO	Inspection Opening
JU	Jump Up
PJC	Plain Junction Cap
P PAN	W.C. outlet through rear wall
RWH	Rain Water Head
S PAN	W.C. outlet through floor.
SC	Stopcock
SD	Sewer Drain
SP	Soil Pipe and Stand Pipe
SV	Stop Valve
SVP	Soil Vent Pipe
SWD	Storm Water Drain
VP	Vent Pipe
WC	Water Closet

Australian Standards Related to House Building

AS 1604.2	Specifications for Preservative Treatment	AS 2699	Built-In Components for Masonary Construction (Wall Ties)
AS 1657	Fixed Platforms, Walkways, Stairways and Ladders - Design, Construction and Installation	AS 2870	Residential slabs and footings
		AS/NZS 2904	Damp proof Courses & Flashings
		AS 3661	Termites
AS 1684.2	Residential Timber Frame Construction (non Cyclonic)	AS 3700	Masonary Code
		AS 3740	Waterproofing of Wet Areas Within Residential Buildings
AS 1684.3	Residential Timber Frame Construction (Cyclonic)		
AS 2047.2	Windows in Buildings	AS 3959	Construction of Buildings in Bushfire-Prone Areas
AS 2699	Built-In Components for Masonary Construction (Connectors & Accessories)	AS 4773.1	Masonry for Small Buildings
		AS 6001	Working Platforms for Housing Construction

Index

Index

House Builders Series

This entirely Australian manual is thoroughly researched in co-operation with the Australian timber, brick and concrete and other relevent associations. It is written in Allan Staines usual easy to comprehend style.

It now has over 660 clear and technically accurate drawings covering each stage of construction. The text and drawings correlate well. The manual covers the foundations and carpentry aspect in detail as well as brick veneer, cavity brick and concrete block systems and weatherboard.

The manual provides current trade practices and hints that help bridge the gap between theory and practice. For these reasons it is an essential class room text.

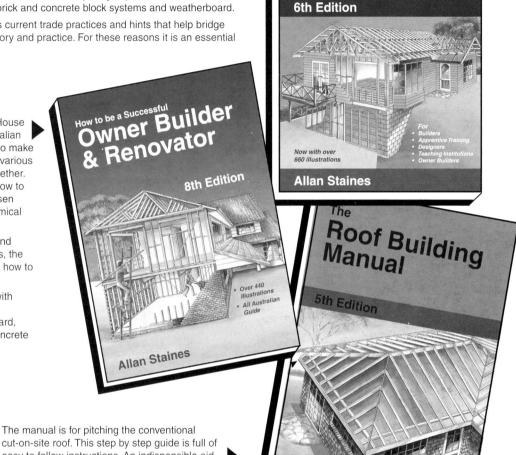

Companion to the Australian House Building Manual. This all Australian title has large scale drawings to make it easy to understand how the various stages and components fit together. The Owner Builder is shown how to plan the house to suit the chosen allotment and the most economical methods of owner building.

You will learn how to identify and correct faulty building practices, the correct trade terms to use and how to order building products.

This manual will provide you with confidence to build a typical Australian home in weatherboard, brick veneer, cavity brick or concrete block.

The manual is for pitching the conventional cut-on-site roof. This step by step guide is full of easy to follow instructions. An indispensible aid for teaching Apprentices, plus quick and easy to use tables and bevels for all roof pitches from 5 degress to 75 degrees. Bevels drawn on the page ready to transfer directly to the bevel tool. A Builder's dream.

Contents of Manuals

Australian House Building Manual
Timber Frame Weatherboard Construction
Brick Veneer Construction
Cavity Brick Construction
Concrete Block Construction
Setting out the Site
Foundations & Footings
Footing & Slab Floors
Timber Floor Framing
Timber Wall Framing
Truss Roof Construction
The installation of fittings such as:
 windows, doors, weatherboards, baths,
 shower recesses and mouldings.
Wet Areas
Flashings, Stair & Handrail construction
Plan Reading
Estimating Quantities
Termite Proofing
Building in Bushfire Areas
Retaining Walls
Adhesives & Sealants
Thin Section Render
Safety & More

How to be a Successful Owner Builder & Renovator
Understanding House Construction
Designing the House
Weatherboard Houses
Brick Veneer Houses
2 Storey Brick Veneer & Weatherboard
Split Level Houses
Double Brick
Concrete Block Masonry
Extensions & Renovations
Extending Brick Veneer Houses
Extending a Weatherboard House
Adding a Verandah or Carport
How to Supervise the Site
Raising a House
Identifying & Correcting Common Problems

The Roof Building Manual
Roofing Basics
Roof Designs
Roofing Members & Where They Fit
Roofing Terms
The Triangle Makes it Easy
The Rafter Length
Marking Out Top Plates & Ridges
Marking Out Rafters
Cutting Out Rafters
Erecting the Roof
Miscellaneous Details
Easy to Follow Rafter Lengths, Tables
& Bevels

ORDERING COUPON OVER PAGE

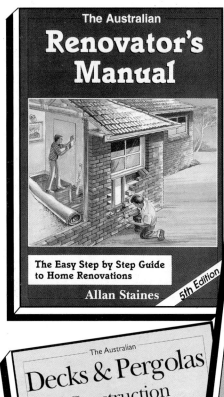

The Australian Renovator's Manual

The Easy Step by Step Guide to Home Renovations

Allan Staines

5th Edition

This book is specifically designed to deal with problems confronted when renovating or altering an existing dwelling. It also shows step-by-step how to accomplish popular renovation projects.

It solves problems such as rising damp, floor and wall cracking, decay in members, small kitchens and bathrooms, leaking shower recesses, and termites. It also describes how to support the roof when making openings in walls.

How to build attic rooms. Installing skylights, guttering, shelving. Replacing existing floor joists, flooring. Fixing sagging roof rafters. Installing french doors. Building carports, enclosing patios and verandahs, installing sliding patio doors, enlarging window openings. Applying Gyprock, restumping, cooling a hot house, attaching laminate and fixing laminate panelling to shower recesses. Wall and floor tiling. Painting. Gives the correct adhesives, sealants and fillers to use and much much more.

A Very Practical Guide to Interior Decorating

The Easy Guide to Decorating Like the Professionals

Carol Staines

2nd Edition

This book will teach you how to successfully decorate your home and achieve that professional decor look. Step-by-step it will show you how to avoid costly mistakes by understanding the key principles of decorating and their relevant application. It will be just like having a professional decorator on-hand to help you with much needed advice.

Chapters cover: How to discover your own decorating style, the principles of design, colour, tonal values, using neutrals, colour combinations to achieve a specific image or atmosphere. How to use pattern and texture, furniture and its placement, using decorating materials, environmentally friendly options, hands-on decorating and decorating to re-sell.

An easy to follow, well illustrated manual.

The Australian Decks & Pergolas Construction Manual

Allan Staines

5th Edition Revised

This all Australian manual is packed with ideas and know how and will suit the Builder, Architect or Home DIY'er. It provides many fresh ideas in colour and easy to follow step-by-step instructions cover every procedure. From designing and drawing an acceptable plan for Councils to constructing the post supports, attaching decks or pergolas to the house, methods of handrailing and lots more. All using proper trade procedures. The easy to follow tables included have been prepared by the Timber Associations - TDA in Sydney and Timber Queensland in Brisbane. These cover footing sizes, posts, bearers, joists, deck fastening, handrailing, pergola bearers, rafters and battens. The manual is an indispensible tool in the hands of Designers and Architects.

Prices Include GST

Take a photocopy of this page and indicate your requirements on the order form. Books are normally return mailed within three days of receiving order.

Qty.	Title	Cost	Total	Enclose a cheque or money order
	Decks & Pergolas Construction Manual	$22.00		Name:
	The Australian House Building Manual	$39.50		
	Successful Owner Builder & Renovator	$37.00		Address:
	The Roof Building Manual	$29.00		
	The Australian Renovators Manual	$27.50		
	A Practical Guide to Interior Decorating	$22.00		
Special Offer 1	Any Four of the above books for	$113.00		
Special Offer 2	Buy Six - One of each title for	$150.00		Postcode:
	Postage and Packaging	$7.00		Phone:
	Payment Enclosed	$		

Please send orders to: Pinedale Press   2 Lethbridge Court
Caloundra QLD 4551 Fax Number (07) 5491 9219
Resellers: Please contact the publisher for reseller Prices. pinedalepress@bigpond.com